Sensors for Ultrasonic NDT in Harsh Environments

Sensors for Ultrasonic NDT in Harsh Environments

Special Issue Editors

Anthony N Sinclair
Robert E. Malkin

MDPI • Basel • Beijing • Wuhan • Barcelona • Belgrade • Manchester • Tokyo • Cluj • Tianjin

Special Issue Editors

Anthony N Sinclair
Department of Mechanical
and Industrial Engineering,
University of Toronto
Canada

Robert E. Malkin
Department of Mechanical
Engineering, University of
Bristol
UK

Editorial Office
MDPI
St. Alban-Anlage 66
4052 Basel, Switzerland

This is a reprint of articles from the Special Issue published online in the open access journal *Sensors* (ISSN 1424-8220) (available at: https://www.mdpi.com/journal/sensors/special_issues/ Sensors_Ultrasonic_NDT).

For citation purposes, cite each article independently as indicated on the article page online and as indicated below:

LastName, A.A.; LastName, B.B.; LastName, C.C. Article Title. *Journal Name* **Year**, *Article Number*, Page Range.

ISBN 978-3-03928-422-1 (Pbk)
ISBN 978-3-03928-423-8 (PDF)

Contents

About the Special Issue Editors

Anthony N Sinclair is a professor in the Department of Mechanical & Industrial Engineering at the University of Toronto. His research specialty is Ultrasonic Nondestructive Evaluation (NDE), with a focus on image enhancement via signal processing, phased arrays, precise measurement of defect size, ultrasonic transducer design, and characterization of material interfaces. His work involves a combination of experimental and numerical modelling techniques, reported in over 200 journal and conference publications, and technical reports. Sponsors of current and past projects have included Ontario Hydro/OPG, NSERC, Pratt & Whitney Canada, NRC Institute for Aerospace Research, Rockwell International, Sigmabond Technologies, Cercast Aluminum, Tower Automotive, Atomic Energy of Canada, DRDC, MITACS, Hatch, Alcan International, ANDEC Manufacturing, Olympus NDT Canada, Advanced Measurement and Analysis Group, Eclipse Scientific, and Groupe Mequaltech. Tony Sinclair was Chair of the Department of Mechanical & Industrial Engineering at the University of Toronto, 2004–2009. He is a past winner of the Faculty Teaching Award for the Faculty of Applied Science & Engineering at U of T. He is on the editorial board of NDT&E International, and has supervised approximately 60 graduate students, postdoctoral fellows, and research associates, plus 90 undergraduate thesis students.

Robert E. Malkin is senior research associate in the Mechanical Engineering department at the University of Bristol. He is also research lead at Ultraleap Ltd. His specialty is advanced ultrasonic imaging, array design and acoustics. His work involves both experimental and numerical simulations of acoustics in both solids and fluids. Between 2014 and 2019, he was lead researcher for a UK-Japan partnership on acoustic imaging of nuclear debris at the Fukushima nuclear power plant. He has had 20+ undergraduate students and 4 post-graduate students. He also lectures for the Nuclear Engineering MSc with a specialty of nuclear accidents. He has consulted for Magnox Ltd, the National Composites Centre, Reckitt Benckiser, Imasonic Ltd, the British Broadcasting Corporation (BBC) and BAE Systems Ltd. He was awarded the British Science Association Award in 2016. He has authored 20 peer reviewed papers and given many international talks (including invited and plenary).

 *sensors*

Editorial

Sensors for Ultrasonic Nondestructive Testing (NDT) in Harsh Environments

Anthony N. Sinclair [1,*] and Robert Malkin [2]

[1] Department of Mechanical and Industrial Engineering, University of Toronto, 5 King's College Road, Toronto, ON M5S 3G8, Canada

[2] Department of Mechanical Engineering, University of Bristol, Bristol BS8 1TR, UK; Rob.Malkin@bristol.ac.uk

* Correspondence: sinclair@mie.utoronto.ca

Received: 23 December 2019; Accepted: 9 January 2020; Published: 14 January 2020

Abstract: In this special issue of *Sensors*, seven peer-reviewed manuscripts appear on the topic of ultrasonic transducer design and operation in harsh environments: elevated temperature, high gamma and neutron fields, or the presence of chemically aggressive species. Motivations for these research and development projects are strongly focused on nuclear power plant inspections (particularly liquid-sodium cooled reactors), and nondestructive testing of high-temperature piping installations. It is anticipated that we may eventually see extensive use of permanently mounted robust transducers for in-service monitoring of petrochemical plants and power generations stations; quality control in manufacturing plants; and primary and secondary process monitoring in the fabrication of engineering materials.

Keywords: ultrasound; nondestructive testing; ultrasonic transducer; high temperature; radiation

1. Introduction

Industrial use of ultrasound for nondestructive testing (NDT) has expanded rapidly over recent years, but has generally been confined to inspections at ambient temperatures, in radiation-free, chemically-benign surroundings. Most of the sensors are based on a lead-zirconate-titanate (PZT) piezoelectric element that has a Curie temperature of a few hundred degrees Celsius, and various plastic components that would not survive temperature excursions above 200 °C. In addition, the degradation of adhesives, and breakdown of various transducer components would threaten such traditional ultrasonic transducers exposed to harsh environments.

In the face of increasingly stringent demands from regulatory bodies that oversee critical industries, and demands from industrial leaders, the demand for ultrasonic transducers that can withstand harsh conditions has risen sharply. Key drivers include:

- On-line, full-power monitoring of crack growth, corrosion, and flow rates in hot piping systems (e.g., nuclear power plants and petrochemical installations).
- Process control and immediate inspection of products on manufacturing production lines (e.g., steel mills and casting plants).
- Nuclear power accident investigation and monitoring as well as nuclear decommissioning inspection/assessment of contaminated materials and structures.
- Corrosion monitoring in chemically aggressive environments (e.g., geothermal piping systems).

There are various mechanisms that have been explored to deal with harsh environments for ultrasonic sensors. These include a selection of materials that resist degradation; use of forced (water) cooling systems; delay lines and wedges that effectively isolate the transducer from the aggressive agent; and air-coupling or electromagnetic acoustic transducers (EMATs) that place a layer of air between the sensor and item under inspection. In this special issue of *Sensors*, seven peer-reviewed manuscripts explore some of the engineering challenges encountered with the design and operation of ultrasonic transducers under these imperfect conditions.

2. Contributions

The first paper by Dhutti et al. [1] from Brunel University UK, focuses on the on-line integrity monitoring of high temperature piping systems, by the use of a piezoelectric wafer active sensor to generate torsional guided waves. As piping temperatures reach up to 600 °C, a gallium phosphate ($GaPO_4$) piezoelectric element was selected, which will resist depoling over extended time periods. Challenges such as differential thermal expansion of transducer components, characterization of material properties, and generation of multiple modes are explored with the aid of a finite element model. Testing conducted for periods of over 1000 h indicate the potential for permanently mounted transducers on piping systems for long term condition monitoring.

The second paper by Bhadwal et al. [2] investigates the problem of coupling together the various components of an ultrasonic transducer that would undergo excursions to temperatures as high as 700 °C. Due to the differential thermal expansion of the piezoelement, backing layer, and wear plate, solid coupling techniques such as brazing would lead to cracking or even total transducer failure. The authors evaluated the feasibility of using dry coupling, but at far lower clamping loads than the 200–500 MPA interfacial pressure often quoted for high dry coupling efficiency. Their transducer design is spring-loaded to maintain a constant inter-layer pressure as the transducer undergoes large temperature excursions; using soft coupling foils at each transducer component interface demonstrated that dry coupling at pressures of 25 MPA can lead to a high signal-to-noise ratio (SNR) and signal amplitude.

The third paper by Pucci et al. [3] originates from the French Alternative Energies and Nuclear energy Commission (CEA) which has a strong program for development of liquid sodium-cooled fast reactors. The ultrasonic sensor is a 12-element array based on the principle of an EMAT array designed for in-service reactor inspection. The paper is concentrated on design details and laboratory testing of prototype transducers; CIVA simulation software is used for simulation and to optimize focusing characteristics. A major advantage of EMAT design is that very accurate control of beam steering and mode generation can be achieved through a judicious choice of array architecture, coil geometry, and magnet arrangement. The inherently weak transduction mechanism of an EMAT is a major challenge, addressed to the extent possible in the design process.

The fourth paper (also from the French CEA) by Le Jeune et al. [4] is again motivated by interest in ultrasonic imaging in in-service sodium-cooled fast reactors. However, the focus here is on optimization of a long-distance imaging system that consists of two linear phased arrays (termed antennas) that are perpendicular to each other. Full Matrix Capture is used for data collection, and then Total Focusing Method is applied to the data for imaging of large areas. CIVA simulations of the system are compared with experimental results of data collected in water; a prototype system for under-sodium trials is currently being manufactured. Extensive plans are described to reduce computation time and improve SNR.

The fifth paper by Saillant et al. [5] also deals with the design of a transducer to be used for inspections conducted in liquid sodium, at temperatures on the order of 200 °C. Although the proposed transducer is based on the same principles of conventional room-temperature probes, the combination of elevated temperature, radiation fields, and a chemically aggressive environment necessitates several design modifications and a selection of appropriate materials. A significant addition is a "wetting layer" to allow ultrasound to pass efficiently between the transducer and liquid sodium. Extensive test result are presented for inspection of machined defects in a welded stainless steel test block.

The sixth paper by Tittmann et al. [6] originates from a long record of research at Penn State University into ultrasonic transducers for use in nuclear reactors. A very thorough review is provided of the piezoelectric materials that could be used under conditions of temperatures over 1000 °C, neutron fluence up to 10^{20} cm^{-2}, and gamma ray dose rates of up to 10^9 Rem/h. Alternative configurations include spray-on transducers and guided wave sensors. Experimental results are presented of transducer performance as a function of irradiation time (both gamma and neutron irradiation) and temperature, for a variety of piezomaterials. This review includes a very extensive bibliography of work done on characterization of piezomaterial properties, and transducer designs for harsh environments.

The last paper of this special issue of *Sensors* comes from the Korean University of Science and technology. Geonwoo Kim et al. [7] are developing an ultrasonic transducer for a very specific task: inspection of pressurized water reactors fuel rods for the detection of infiltrated water (akin to fuel rod failure). A high-sensitivity Pb(Mg$_{1/3}$Nb$_{2/3}$)O$_3$-PbTiO$_3$ single crystal is selected as the active piezoelement, as it can generate stronger signals than PZT-based transducers, in this highly attenuative inspection environment. Extensive results are reported on the effect of neutron irradiation on this material, to show that it is suitable for an in-reactor environment. Transducer design details, backed up by a one-dimensional Krimholtz Leedom Matthae (KLM) model of performance, are presented, for a sensor with central frequency of 5 MHz.

Author Contributions: This editorial was drafted by A.N.S., and edited by R.M. All authors have read and agreed to the published version of the manuscript.

Funding: This research received no external funding.

Conflicts of Interest: The authors declare no conflict of interest.

References

1. Dhutti, A.; Tumin, S.A.; Balachandran, W.; Kanfoud, J.; Gan, T.-H. Development of Ultrasonic Guided Wave Transducer for Monitoring of High Temperature Pipelines. *Sensors* **2019**, *19*, 5443. [CrossRef] [PubMed]
2. Bhadwal, N.; Torabi Milani, M.; Coyle, T.; Sinclair, A. Dry Coupling of Ultrasonic Transducer Components for High Temperature Applications. *Sensors* **2019**, *19*, 5383. [CrossRef] [PubMed]
3. Pucci, L.; Raillon, R.; Taupin, L.; Baqué, F. Design of a Phased Array EMAT for Inspection Applications in Liquid Sodium. *Sensors* **2019**, *19*, 4460. [CrossRef] [PubMed]
4. Le Jeune, L.; Raillon, R.; Toullelan, G.; Baqué, F.; Taupin, L. 2D Ultrasonic Antenna System for Imaging in Liquid Sodium. *Sensors* **2019**, *19*, 4334. [CrossRef] [PubMed]
5. Saillant, J.-F.; Marlier, R.; Navacchia, F.; Baqué, F. Ultrasonic Transducer for Non-Destructive Testing of Structures Immersed in Liquid Sodium at 200 °C. *Sensors* **2019**, *19*, 4156. [CrossRef] [PubMed]
6. Tittmann, B.R.; Batista, C.F.; Trivedi, Y.P.; Lissenden, C.J., III; Reinhardt, B.T. State-of-the-Art and Practical Guide to Ultrasonic Transducers for Harsh Environments Including Temperatures above 2120 °F (1000 °C) and Neutron Flux above 10^{13} n/cm^2. *Sensors* **2019**, *19*, 4755. [CrossRef] [PubMed]
7. Kim, G.; Choi, N.; Kim, Y.-I.; Kim, K.-B. Pb(Mg$_{1/3}$Nb$_{2/3}$)-PbTiO$_3$-Based Ultrasonic Transducer for Detecting Infiltrated Water in Pressurized Water Reactor Fuel Rods. *Sensors* **2019**, *19*, 2662. [CrossRef] [PubMed]

Article

Development of Ultrasonic Guided Wave Transducer for Monitoring of High Temperature Pipelines

Anurag Dhutti [1,*], Saiful Asmin Tumin [1], Wamadeva Balachandran [1], Jamil Kanfoud [1] and Tat-Hean Gan [1,2,*]

[1]   Brunel University London, Kingston Ln, Uxbridge, Middlesex UB8 3PH, UK;
      sword_73@hotmail.com (S.A.T.); wamadeva.balachandran@brunel.ac.uk (W.B.);
      jamil.kanfoud@brunel.ac.uk (J.K.)
[2]   TWI Lt, Granta Park, Cambridge CB21 6AL, UK
*   Correspondence: anurag.dhutti@brunel.ac.uk (A.D.); tat-hean.gan@brunel.ac.uk (T.-H.G.);
    Tel.: +44-7938-16-3560 (A.D.); +44-12-2389-9000 (T.-H.G.)

Received: 16 November 2019; Accepted: 8 December 2019; Published: 10 December 2019

Abstract: High-temperature (HT) ultrasonic transducers are of increasing interest for structural health monitoring (SHM) of structures operating in harsh environments. This article focuses on the development of an HT piezoelectric wafer active sensor (HT-PWAS) for SHM of HT pipelines using ultrasonic guided waves. The PWAS was fabricated using Y-cut gallium phosphate ($GaPO_4$) to produce a torsional guided wave mode on pipes operating at temperatures up to 600 °C. A number of confidence-building tests on the PWAS were carried out. HT electromechanical impedance (EMI) spectroscopy was performed to characterise piezoelectric properties at elevated temperatures and over long periods of time (>1000 h). Laser Doppler vibrometry (LDV) was used to verify the modes of vibration. A finite element model of $GaPO_4$ PWAS was developed to model the electromechanical behaviour of the PWAS and the effect of increasing temperatures, and it was validated using EMI and LDV experimental data. This study demonstrates the application of $GaPO_4$ for guided-wave SHM of pipelines and presents a model that can be used to evaluate different transducer designs for HT applications.

Keywords: gallium phosphate; piezoelectric wafer active sensor; thickness shear; high-temperature monitoring; ultrasonic; guided wave; structural health monitoring

---

1. Introduction

There is an increasing emphasis on the development of structural health monitoring (SHM) systems using in situ sensors that can inform an operator about the health of their critical infrastructure, develop damage and estimate the remaining useful life. Piezoelectric wafer active sensors (PWASs) are increasingly used for many SHM applications, as they are lightweight, inexpensive and can be permanently installed on structures. They have been successfully demonstrated on various critical components and substructures in many industrial applications [1]. PWASs can be used for damage detection in three different ways: guided-wave ultrasonics [2], high-frequency modal sensing [3] and acoustic emission passive detection [4]. SHM applications for assets subjected to an extreme operational environment such as high temperature (HT) are in great demand to monitor critical components in gas turbines and nuclear power industry. This has led to rapid developments in the field of HT piezoelectric sensing. Several accelerometers, surface acoustic wave sensors, ultrasonic transducers and pressure sensors for temperatures up to 1250 °C have been reported [5].

High-temperature pipelines (HTPs) are critical components in nuclear power plants, providing connections between feed water pumps, a boiler or heat exchanger and a turbine. HTPs are operating at extreme temperatures up to around 600 °C [6], high pressures and subjected to continuous cyclic loadings. This can lead to the development of creep-fatigue- [7], corrosion- [8] and stress corrosion cracking-type [9] defects. Currently, they undergo periodic inspections during maintenance outages, but new technologies are being sought after for in situ monitoring and early warnings of material degradation [10]. A number of SHM technologies have been developed for pipelines [11]. A suitable solution is demonstrated by the ultrasonic guided wave (UGW) technique, as it provides full cross-sectional area coverage for a range of tens of meters from a single location. Rings of shear transducers are used to transmit unidirectional and axially symmetric ultrasonic wave modes [12] in a frequency range of 20–100 kHz, and they can detect the presence of structural defects such as cracks and corrosion by analysing reflected waves. Commercial products such as gPIMS [13] are now available for SHM of pipelines but are limited for continuous operation at temperatures below ~100 °C. To enable guided-wave SHM for HTPs, piezoelectric transducers that can perform reliably at HT are required.

2. High-Temperature Ultrasonic-Guided-Wave Transducers

Several transduction technologies based on piezoelectric, electromagnetic-acoustic and magnetostrictive principles have been developed for the generation and the detection of UGWs. For application at HT, piezoelectric sensing is the most promising technique due to its thermal stability and reliability [5]. A thickness-shear (TS) piezoelectric transducer consists of five main components: a face plate, a piezoelectric element, a backing/damping block, a casing and electrical connections, as illustrated in Figure 1. The electrical connection allows for the transfer of an electrical pulse to and from the piezoelectric element. The casing provides mechanical support for the whole transducer as well as electrical shielding for the piezoelectric element. The damping or backing block damps the vibration of the piezoelectric element, after the excitation pulse is transmitted. The piezoelectric element generates a mechanical signal from an electrical input or, conversely, an electrical signal from a mechanical input, enabling transmission and reception of ultrasonic signals, respectively. The face plate protects the fragile piezoelectric element and is used for mechanical impedance matching.

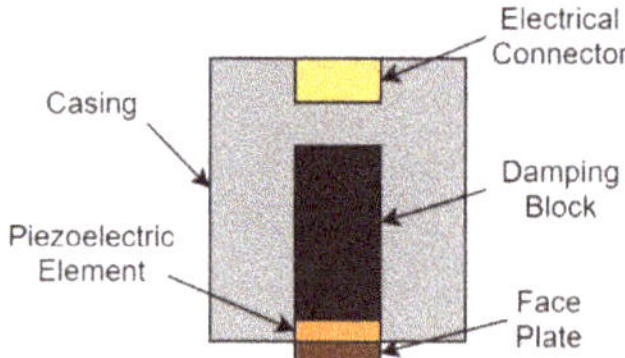

Figure 1. Construction of a generic piezoelectric transducer.

In pipes, there is a large number of wave modes present in the frequency range of interest, and a transducer can excite all the modes that exist within its frequency bandwidth. This can complicate the data and make its interpretation challenging. To simplify the guided wave response, transducers are designed to minimise the excitation of unwanted modes and achieve a linear frequency response in the region of operation to avoid mode coupling, frequency jumps and performance dips. For the HT SHM application, transducers must also provide sufficient ultrasonic outputs at the target temperature and sustain their electromechanical response over prolonged periods of time to achieve the desired defect sensitivity of the SHM system. This requires careful selection of piezoelectric material, understanding of its thermal characteristics and compatibility with other transducer subcomponents.

2.1. High-Temperature Piezoelectric Materials

Piezoelectric materials are characterised by a number of interrelated coefficients standardised by the IEEE [14,15], which define their piezoelectric, dielectric and elastic behaviour. These coefficients are used to describe the coupling between structural and electrical behaviour of the material. It can be expressed as a relation between the material stress and its permittivity at a constant stress (stress charge form, Equation (1)) or as a relation between the material strain and its permittivity at a constant strain (strain charge form, Equation (2)).

$$\text{Stress charge form}: \ T = c^E S - e^T E, \ D = eS + \epsilon^S E \tag{1}$$

$$\text{Strain charge form}: \ S = s^E T + d^T E, \ D = dT + \epsilon^T E \tag{2}$$

where E is the electric field, D is the electric displacement, S is the mechanical strain, and T is the stress. The material parameters s^E, d and ϵ^T are material compliance, coupling property and relative permittivity at a constant stress, respectively, and they are tensors of ranks 4, 3 and 2, respectively, and represented using the abbreviated subscript Voigt notation.

A useful guide for the selection of an appropriate piezoelectric material is its figure of merit (FOM), which is defined based on a specific application [16]. A key FOM for an HT transducer is the product of piezoelectric charge coefficient d and voltage coefficient g, where a higher value gives a higher electromechanical coupling coefficient, k, which is defined as the ratio of stored electrical energy to applied mechanical energy and vice versa.

Several new high-performance HT piezoelectric materials have been developed [17–19], and they can be broadly categorized into two types: ferroelectric polycrystalline (such as lead zirconate titanate ($PbZrTiO_3$ and barium titanate ($BaTiO_3$)) and single crystals (such as lithium niobate ($LiNbO_3$) and quartz (SiO_2)). Ferroelectric materials possess high FOMs (e.g., $k = 0.7$ for $PbZrTiO_3$) and are the most commonly used piezoelectric materials for ultrasonic transducers. However, they are limited to operate at temperatures below their Curie temperatures (T_c). Beyond T_c, they depolarize, i.e., their ferroelectric phase transforms into a high-symmetry nonferroelectric phase, and are no longer piezoelectric. Ferroelectric materials at temperatures exceeding $0.5T_c$ [20] can undergo accelerated thermal ageing and property degradation. They also exhibit pyroelectric behaviour, which can lead to instability of their electromechanical properties at HT. Nonferroelectric single crystal materials are being increasingly explored, as they do not undergo phase transition or domain-related aging behaviour, have high resistivity and low losses and can provide excellent thermal stability. Various types of single crystals have been investigated for HT applications, and monoclinic-type BiB_3O_6 (BIBO) and langasite-type $Ca_3TaGa_3Si_2O_{14}$ (CTGS) trigonal crystals have shown stable piezoelectric properties at temperatures up to 700 and 900 °C, respectively [21].

Gallium orthophosphate ($GaPO_4$) was selected for this study, as it offers high electrical resistivity, low acoustic losses, no pyroelectricity and thermally stable properties (constant value of d_{11} at temperatures up to 700 °C [22]), which makes it a suitable candidate for HT UGW transducers. $GaPO_4$ crystals belong to the trigonal 32-point group and can be synthesised using a hydrothermal process. The AVL (now Piezocryst Advanced Sensorics GmbH) developed the first commercial product GM12D, which was an uncooled pressure sensor for combustion engines, and since then, $GaPO_4$ has been explored for a large field of HT resonator applications including temperature-compensated cuts for bulk acoustic wave (BAW) [23], surface acoustic wave (SAW) [24] and microbalance [25] operating at temperatures up to 720 °C. $GaPO_4$ resonators can be obtained in different crystal cuts and depending on crystal cut angles, and their sensitivity, resolution and linearity differ in a wide range. Singly rotated Y-cut $GaPO_4$ resonators with different angles have been investigated [26]. Y $-84°$ was shown to have a very high Q factor and a high drive level dependency appropriate for pressure sensing applications, whilst Y $+27°$ displayed a linear resonance frequency dependence on temperature [27], which makes it suitable for temperature sensing. An X-cut resonator was investigated for phased arrays and displayed encouraging results for SHM applications [28].

2.2. High-Temperature Transducer Design

The selection of transducer passive materials including electrodes, matching layers, backing materials, connecting wires and adhesives also requires careful consideration. The coefficient of thermal expansion (CTE) of the piezoelectric material should be compatible with the structure and the materials within the particular transducer assembly. Excessive differential thermal expansion can lead to permanent damage to the transducer assembly or thermal stresses within the piezoelectric material, leading to the generation of unwanted modes of vibration. Bonding of different materials within the transducer assembly is also challenging, as the temperature required for bonding must be below $T_c/2$ and pressure used in bonding could crack the piezoelectric material. The bond must also remain strong at the operating temperature of the device and be elastic in order to be able to change shape, as the piezoelectric deforms and the other parts of the transducer expand or contract due to changes in temperature.

Common methods of bonding include the use of HT epoxy, which is a thermosetting polymer used as a glue. Additionally, glass solder can be used at temperatures up to 500 °C, as for higher temperatures, it reacts chemically with other components. Other methods include regular soldering, diffusion bonding, ultrasonic welding, cementing, sol-gel and vacuum brazing. A comprehensive review of these bonding techniques and designs of ultrasonic transducers at HT can be found in the literature [17]. Unlike bulk transducers that require lower internal damping to achieve resonant transducers, piezoelectric materials must be damped heavily to produce broadband, guided-wave transducers.

2.3. Thickness-Shear Piezoelectric Material Characterisation

It is critical to characterise piezoelectric properties for the transducer design and modelling and also to analyse their performance for the application under study. Not all the parameters are usually available from datasheets, the real values may deviate from specified values due to electrodes and polarisation, or they might not be available for the range of the temperature of interest. The material properties can be derived from their electromechanical impedance (EMI) measurements in accordance with BS EN 50324-2:2002 [29]. The expression of the EMI of a plate vibrating in a TS mode as a function of frequency can be shown as:

$$z(\omega) = \frac{t}{i\omega\epsilon_{22}^S A}\left[1 - k^2\frac{\tan\left(\frac{\omega}{4f_a}\right)}{\left(\frac{\omega}{4f_a}\right)}\right],\tag{3}$$

where t is the plate of thickness t and A is the electrode area. The EMI response represents the dynamic behaviour of the piezoelectric material, when an AC voltage is applied. An example of the EMI response of a commonly used PZT material is presented in Figure 2.

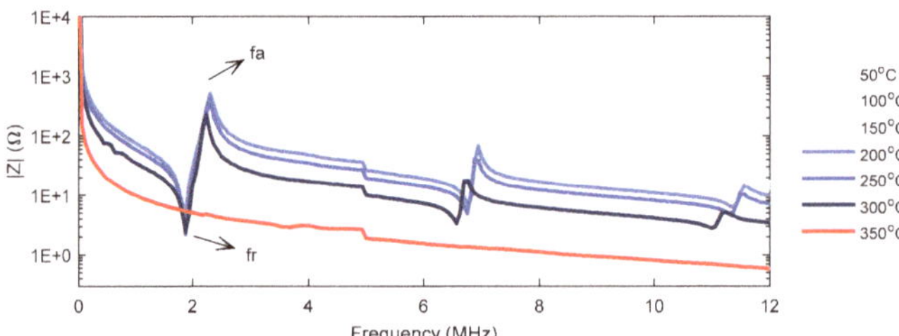

Figure 2. Example of high-temperature electromechanical impedance (EMI) response from a Lead zirconate titanate (PZT) plate.

The electromechanical resonances are associated with the reaction of a ceramic body to an ultrasound. EMI versus frequency plots can be presented on a logarithmic scale to emphasise weaker excitable modes including a harmonics, spurious or unwanted mode. For piezoelectric materials in class 32, the TS mode of a vibration can be obtained using Y-cut plates with electrode surfaces on a plane normal to the x_2 direction. For the TS mode, the eigenfrequency equation for the harmonics is given as:

$$f_n = n\frac{v}{2h}, \quad v = \sqrt{\frac{c_{66}^D}{\rho}},$$
(4)

where f_n is the frequency of the nth harmonic, h is the thickness of the plate, and v is the shear acoustic wave speed, which is related to the material density ρ and the elasticity constant c_{66}^D. The characteristic resonance frequencies of the TS mode can be used to compute the electromechanical coupling factor k_{26} by:

$$k_{26}^2 = \frac{\pi}{2}\frac{f_r}{f_a}\tan\left(\frac{\pi}{2}\frac{\Delta f}{f_a}\right),$$
(5)

where f_r and f_a are the resonance and antiresonance frequencies, respectively, as annotated on the EMI measurements in Figure 2. Other relevant material constants for the TS mode include the elastic constants c_{66}^E and S_{66}^E, the piezoelectric constants e_{26} and d_{26} and the dielectric constant ϵ_{22}^S. These parameters are related to the electromechanical coupling factor k_{26}, which can be written as:

$$\frac{k_{26}^2}{1 - k_{26}^2} = \frac{e_{26}^2}{\epsilon_{22}^S c_{66}^E}.$$
(6)

The elastic constants can be derived using EMI characteristic frequencies, material density and plate thickness, which can be expressed as:

$$c_{66}^E = \left(1 - k_{26}^2\right)\rho f_p t^2, \quad c_{66}^D = 4\rho(t f_r)^2,$$
(7)

$$S_{66}^E = \frac{1}{4\rho t^2 f_a^2\left(1 - k_{26}^2\right)}.$$
(8)

The temperature coefficients of these elastic, piezoelectric and dielectric constants are important for predicting the variation of the frequency response of practical transducers with temperature. The temperature coefficients of the material properties are defined as:

$$T(P)^{(n)} = \frac{1}{P_0 n!}\left(\frac{\delta^n P}{\delta\theta^n}\right)_{\theta = T_0}, \quad P(T) = P(T_0) + \sum_{n=1}^{3} T(P)^{(n)}\cdot(T - T_0)^n,$$
(9)

where $T(P)^{(n)}$ is the nth-order temperature coefficient of the material property P, T_0 is the reference temperature (usually 25 °C), and the property $P(T)$ at temperature T can be calculated using these coefficients.

3. Materials and Methods

In this study, a $GaPO_4$ plate was examined for the HT guided-wave ultrasonic transducer development. A finite element model was developed to simulate the electromechanical response of the PWAS. The effects of temperature and material degradation on the electromechanical response were evaluated using the model. The modes of vibration within the frequency range of interest for the UGW were also analysed to evaluate suitability and performance as a TS-mode transducer. Both the simulated EMI and the vibration response were validated through experiments. The model development and the experimental setups were described in this section. The methodology followed in this study is highlighted in Figure 3, which shows how the results will be used for a complete transducer design.

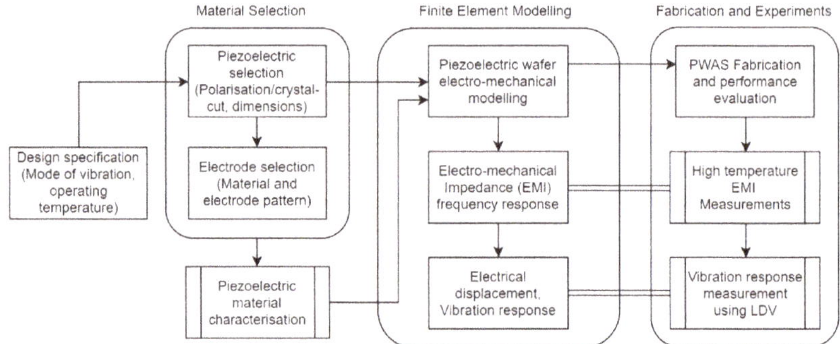

Figure 3. Flow chart showing the methodology followed in this study for transducer design, modelling and validation. Experiments are shown as processes linked to the validation of models.

3.1. GaPO$_4$ Piezoelectric Wafer Transducer

GaPO$_4$ resonator plates were obtained by Piezocryst Advanced Sensorics GmBH in a Y-cut (YXl)0° configuration to achieve the desired thickness shear response. The dimensions of the resonator plates were y = 0.5 mm, x = 13 mm and z = 3 mm. These dimensions were chosen to achieve a clean TS mode of vibration for accurate k measurements, as plates with an L/t ratio of <10 lead to a coupling of the TS mode with additional miscellaneous modes [30]. Platinum (Pt) was chosen as the electrode material, as it has excellent conductive properties, resists oxidation and has demonstrated HT performance with no degradation at temperatures up to 650 °C [31,32]. No electrode-matching layer was applied between the two materials, as the CTE of Pt (9×10^{-6}/°C) matches well with the CTE of GaPO$_4$ (12.78×10^{-6}/°C). Pt electrodes with a thickness of 100 nm were applied using the magnetron sputtering physical vapour deposition (PVD) technique in a wrap-around configuration. The electrode for the ground connection was applied on one of the two large faces, its adjacent end face and part of the second large face, as shown in Figure 4. The electrode for the input voltage was applied on the second large face with a 1 mm gap from the ground electrode, which was achieved by applying a micro-masking tape before sputtering.

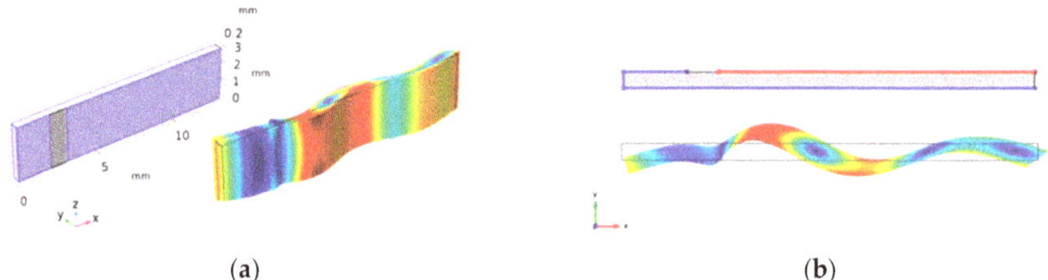

(a)   (b)

Figure 4. Y-cut GaPO$_4$ model geometries and wrap-around electrodes used for 3D (**a**) and 2D (**b**) simulations.

3.2. Modelling of the GaPO$_4$ Transducer

A finite element model of the GaPO$_4$ PWAS was developed using the COMSOL Multiphysics (version 5.3a) software package to simulate its piezoelectric behaviour. The piezoelectric device module of COMSOL related the strain and the electric field using Equations (1) and (2). The objectives of this model were to study the impedance resonance spectrum and the modes of vibration within the frequency range of interest for the guided-wave ultrasonic application.

Due to the anisotropic nature of GaPO$_4$ (trigonal symmetry), it is necessary to have a full 3D model for accurate estimations, as resonance frequency is highly dependent on crystal orientation (cut angle). However, for a 0°-rotation, Y-cut PWAS, a simpler 2D geometry could be used. Plane strain 2D approximation was used in the model. The results from both the 2D and 3D models were compared, and the TS-mode results were the same. The model geometries and a comparison of the simulated TS-mode shapes are shown in Figure 4. In both cases, the thickness of the PWAS was aligned along the y-axis to represent the Y-cut. All the FEA studies in this paper were performed in two dimensions to reduce computation requirements. The dimension, width, was also included as an out-of-plane thickness in the model.

Representative electrical boundary conditions were applied to the PWAS. The 2D geometry in Figure 4 highlights the edges, where the electrical boundary conditions for the ground (blue) and AC terminal (red) voltages were applied. The zero charge condition was applied to all the other edges. Previous studies on the mass loading effect of Pt electrodes have shown that electrodes do not have a significant effect on transducers, where the wavelength is much greater than the electrode thickness [33]. Therefore, in this model, the electrodes were assumed equipotential, i.e., ideal conductor with no atomic mass.

The effects of temperature and material degradation were investigated by introducing temperature-dependent material properties of $GaPO_4$. The elasticity, piezoelectric and dielectric permittivity matrices and their temperature coefficients were acquired from the material datasheets and the literature [34]. The density at different temperatures was calculated using the following relation:

$$\rho = \frac{m}{lwt} = \frac{m}{l_0 w_0 t_0 \left(1 + \frac{\Delta l}{l_0}\right)\left(1 + \frac{\Delta w}{w_0}\right)\left(1 + \frac{\Delta t}{t_0}\right)} = \frac{\rho_0}{\left(1 + \frac{\Delta l}{l_0}\right)\left(1 + \frac{\Delta w}{w_0}\right)\left(1 + \frac{\Delta t}{t_0}\right)}, \tag{10}$$

where ρ_0 is the density at room temperature, $\Delta l / l_0$, $\Delta w / w_0$, and $\Delta t / t_0$ were calculated based on the CTEs provided in the datasheet using the following relation. The CTEs α_{ij} of $GaPO_4$ is a second-rank tensor with only two independent coefficients α_{11} and α_{33}. The temperature coefficients for CTEs and elastic properties were used to compute their respective HT values for the model using Equation (8). The HT material properties are presented in Figure 5.

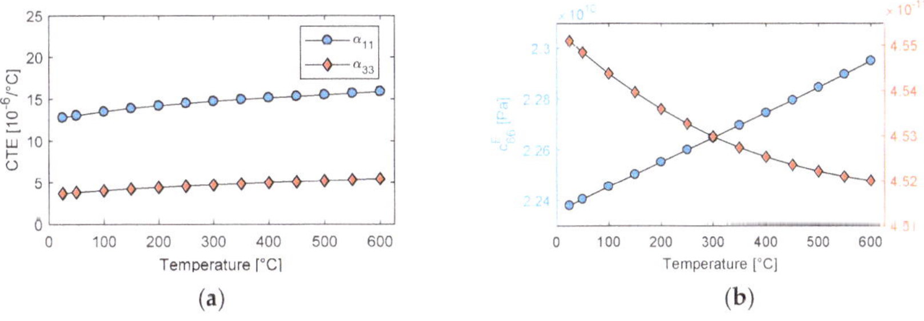

Figure 5. Temperature response of (**a**) coefficients of thermal expansion (CTEs) and (**b**) elastic coefficients, calculated using temperature coefficients from the literature.

A fine mapped mesh was utilised in order to achieve a uniform mesh shape throughout the geometry. Previous modelling studies have shown that, for the TS mode, the number of mesh elements along the thickness can have a significant effect on the simulated resonance frequency [35]. A mesh analysis was performed by analysing the characteristic frequencies from the second resonance with an increasing number of mesh elements. Frequencies from the second resonance were chosen, as they are more significantly affected by the thickness compared to the fundamental resonance. From the results in Figure 6, it can be seen that frequency convergence was achieved with 12 mesh elements along the thickness. This related to 14 elements per wavelength at the highest frequency of interest. A mapped mesh with a size of 0.04 mm × 0.04 mm (length × width) was used in all models, which resulted in 3900 mesh elements, covering an area of 6.5 mm^2.

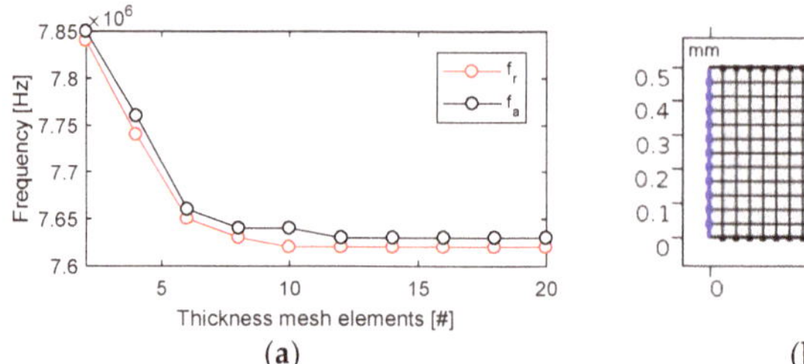

Figure 6. Mesh analysis results showing the change in frequency with an increasing number of mesh elements along the thickness (**a**) and the chosen mesh in the model (**b**).

The model with the optimised mesh was then used to perform two different frequency-domain studies. The first study was conducted within a frequency range of 40 to 12 MHz, to simulate EMI spectra and analyse TS-mode resonance frequencies. The EMI spectrum was extracted using the global evaluation function on the inverse of the admittance parameter (es.Y11) in COMSOL. The second study was performed to evaluate the electrical displacement patterns for the transducer vibrational analysis within a frequency range of 10 to 150 kHz.

3.3. Electromechanical Impedance Characterisation

EMI measurements were performed on the GaPO$_4$ plates described in Section 3.1. The measurements were performed using a purpose-built test rig [36], which includes a Carbolite LHT6-30 Furnace, and the sample was connected to an Agilent 4294a Impedance/Gain-Phase Analyser through an Agilent 16048A fixture. This test rig was used to evaluate different HT piezoelectric materials, and a comparison of their HT TS-mode properties was reported [37]. The exact dimensions of the samples were measured at an accuracy of 0.01 mm using a digital calliper. Impedance measurements were carried out initially at room temperature and subsequently at temperatures increased with 50 °C intervals up to 600 °C. Before each measurement, the jig was recalibrated by performing a short/open circuit compensation to suppress any parasitic impedance. The impedance spectra were recorded over a frequency range of 40 to 12 MHz with an increment of 15 kHz.

The TS-mode properties were calculated using Equations (3)–(6), where the measured dimensions, the densities of the samples and the frequencies F_r and F_a were used. For HT values, the properties were calculated using HT dimensions and density. The HT dimensions were calculated by applying Equation (8), and the density at each temperature was calculated using Equations (10). Each of these properties was calculated at several temperatures, and their temperature coefficients were found using curve fitting and the relation defined in Equation (8). The properties and their temperature coefficients were averaged for five samples.

3.4. Vibrational Response Measurement

The ultrasonic vibrational response of the GaPO$_4$ PWAS was measured using laser Doppler vibrometry (LDV); the experimental setup is shown in Figure 7. The PWAS were connected to a pulser-receiver unit by soldering thin wires on the Pt-coated surface of the GaPO$_4$ plates. A Teletest Focus+ [38] pulser-receiver was used to provide a chirp excitation signal with a frequency range of 10–150 kHz, and a Polytec PSV-400-3D-M scanning laser vibrometer was used to obtain 3D LDV measurements of the GaPO$_4$ PWAS surface. The laser beams of the 3D scanner were focused on the surface of the element. 2D and 3D calibration procedures were performed, before the scan points were set to cover the entire surface area of the samples. The collected 3D displacement measurements from all the scan points were averaged and transformed into the frequency domain to evaluate the displacement spectrum.

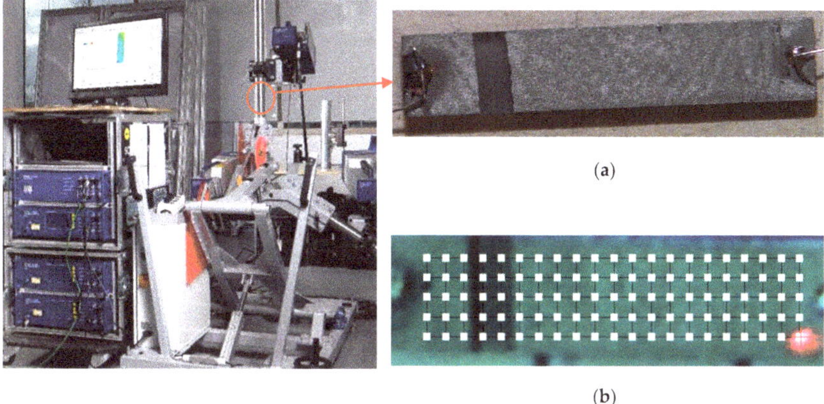

Figure 7. 3D laser Doppler vibrometry experimental setup for measuring the vibrational response of the GaPO$_4$ piezoelectric wafer active sensor (PWAS) specimen (**a**) showing the scan points used for 3D measurements (**b**).

4. Results

4.1. Electromechanical Impedance Response

A comparison of the EMI measurements and the modelled responses of the Y-cut GaPO$_4$ plate is shown in Figure 8. Two TS-mode resonances around 2.5 and 7.6 MHz can be observed and were consistent with the frequencies calculated based on Equation (4). A comparison of the calculated, measured and modelled resonances is provided in Table 1. The modelled phase of the impedance was always $\pm\pi/2$ as in the model, and the resonator was assumed lossless. The coupling to high overtone contours and undesired modes can be seen in the antiresonance frequency of the fundamental resonance, which was also simulated by the model. These unwanted modes usually had lower frequencies, and their influence became smaller for the overtones, as seen in the second overtone.

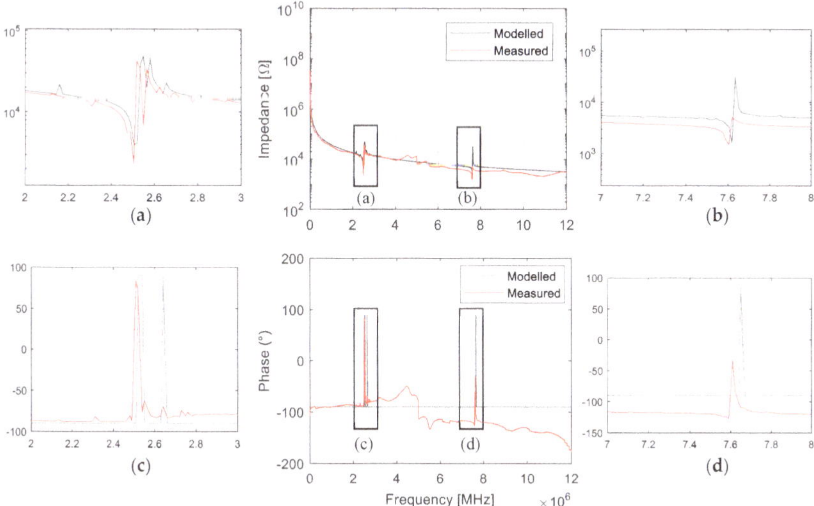

Figure 8. Comparison of the EMI measurements and the modelled responses highlighting impedance amplitude (**a**) and phase (**b**) of the fundamental TS-mode resonance and impedance amplitude (**c**) and phase (**d**) of the overtone.

A slight frequency shift in the measured EMI that affected both resonance and antiresonance frequencies was observed. This was the result of the mass loading effect from the deposited resonator electrode and the connecting wires, which were not included in the model. This is confirmed by the higher disparity in the second resonance, as mass loading is more significant for high overtone resonances, as shown in previous studies [33].

Table 1. Comparison of theoretical thickness-shear (TS)-mode frequencies between measurements and simulation.

TS Mode	Frequency (MHz)			Relative Error (%)
	Calculated	Measured	Simulated	
Fundamental (n = 1)	2.53	2.52	2.51	0.4
Overtone (n = 3)	7.59	7.61	7.62	0.13

The measured and simulated EMI responses at increasing temperatures, along with its effects on resonance frequency and capacitance, are shown in Figure 9. The measured impedance amplitude at HT was reduced by a factor of 10 compared to that of the response at ambient temperature, and also no mode coupling was observed around the antiresonance frequency. This resulted from the HT conductive adhesive used to attach electrical connection to the PWAS. For the HT models, mechanical damping with an isotropic structural loss factor of 0.01 was added to obtain a comparable frequency bandwidth with the measured response.

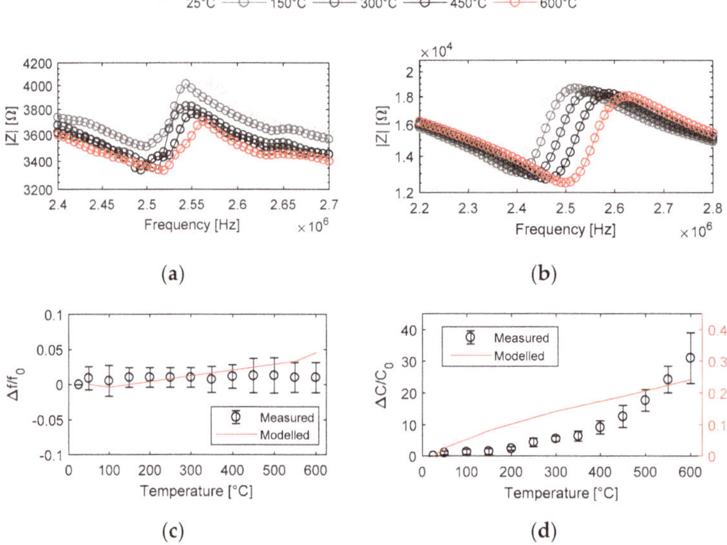

Figure 9. Comparison of EMI response at high temperature: (**a**) measured EMI amplitude; (**b**) modelled EMI amplitude; (**c**) relative change of frequency at temperatures up to 600 °C; and (**d**) relative change in capacitance at temperatures up to 600 °C.

With an increase in temperature, a decrease in impedance amplitude and number of antiresonance peaks can be seen. This related to the increase in permittivity and capacitance of the material [39]. Both the measured and simulated EMI responses at HT are in good correlation, which can be seen by a relative shift in resonance and antiresonance frequencies with increasing temperature. The difference in impedance amplitude was associated with the electrical connections within the measurement setup and their parasitic impedance. The temperature coefficient of the resonance frequency predicted by the model was $27.89 \times 10^{-6}/°C$, which matched well with the value of $27.3 \times 10^{-6}/°C$ calculated from the measured response with an error of less than 2.2%.

The capacitance data from the model was extracted by computing an integral of the total electric energy on the electrode edge. The increase in capacitance at higher temperatures was observed in both the measured and simulated response. The change in measured capacitance was around 100 times more than that in simulated capacitance. This is due to the combined effect of dielectric losses in $GaPO_4$ and parasitic capacitance from the electrical connections and the measurement setup.

4.2. TS-Mode Coefficients of GaPO₄

The TS-mode coefficients were derived from the EMI measurements using Equations (6)–(8). The calculated coefficients were compared with those reported in the literature, and the relative errors are listed in Table 2. The measured elastic coefficients are in good agreement with those reported previously [40]. There was a considerably larger difference in the measured k_{26}. This is because of the difference in PWAS geometry, as the previous study used a circular plate with gold electrodes [40]. The deviation was attributed to the electric field concentration at electrode corners in a rectangular design, which was reduced in circular designs.

Table 2. GaPO₄ TS-mode properties at room temperature.

Property	Units	Ref [41]	Measured	Modelled	Relative Error (%)
C_{66}^E	GPa	22.38	21.76	-	2.85
C_{66}^D	GPa	23.28	22.36	-	3.95
S_{66}^E	$10^{-12}\ m^2N^{-1}$	45.51	45.33	-	0.4
k_{26}	-	0.183–0.192 [1]	0.164	0.169	7.65

[1] The reported value is for a different geometry and a different electrode configuration.

The variation between the measured and modelled electromechanical coupling factors k_{26} was less than 3%. This showed that the model can be accurately used to predict the electromechanical resonances and the associated constitutive piezoelectric properties that indicate its performance as an ultrasonic transducer.

The material properties calculated from the HT EMI measurements in Figure 10 were compared with those calculated using the first-, second- and third-order temperature coefficients provided in the datasheet. The temperature dependence of elastic constants exhibited a stable state at temperatures up to 600 °C with a variation of less than 3%. The measured k_{26} was extremely consistent, showing a less than 1% variation at temperatures up to 600 °C. This demonstrated excellent temperature-independent behaviour.

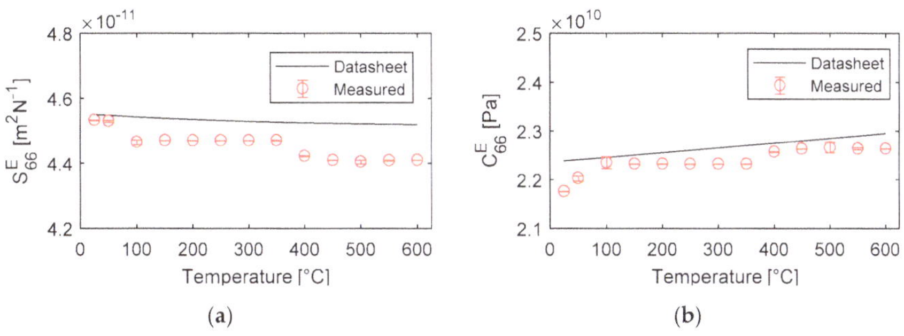

Figure 10. Assessment of the measured elastic constants (**a**) S^E and (**b**) C^E with those calculated using the temperature coefficients provided in the literature and the datasheets.

The first-order temperature coefficients were derived from the measured HT properties using Equation (8). The derived coefficients were compared with those available in the literature and are presented in Table 3. A relative error of approximately 3% was observed for the three key FOMs for the TS-mode transducer.

Table 3. Temperature coefficients of the TS-mode GaPO$_4$ elastic and piezoelectric properties.

Property	$Tp_{ij}^{(1)}$ $(10^{-6}\ K^{-1})$		Relative Error (%)
	References [40,41]	Measured	
C_{66}^{E}	44.9	46.4	3.23
S_{66}^{E}	−26.9	−27.8	3.34
f_r	19.9–35.7 [1]	27.3	1.8 [2]
k_{26}	452	469	3.62

[1] Reported values are with and without correction from stray capacitance [41]. [2] Error calculated based on the average of the two reported values.

The variation in the properties for the estimation of the transducer FOMs was recorded at 600 °C for a period of 1000 hours. The results in Figure 11 show a consistent response from the beginning, with variations of less than 1% in k_{26} and 1.5% in its s_{66}^{E} and c_{66}^{E}, which demonstrated an excellent thermal aging behaviour of TS-mode properties, proving its suitability for SHM applications.

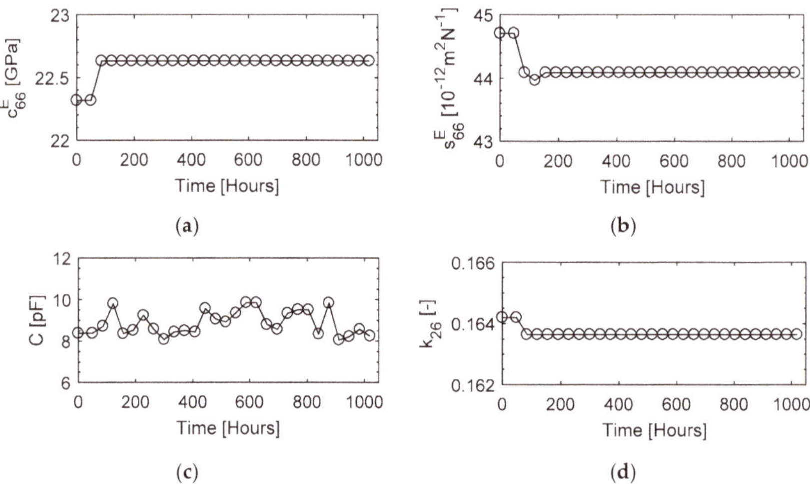

Figure 11. Thermal aging response of the TS-mode elastic coefficients (**a**) C^E, (**b**) S^E; (**c**) capacitance; and (**d**) piezoelectric coefficient k$_{26}$ measured at 600 °C for over 1000 h.

4.3. Vibration Analysis

A comparison of the vibration response measured using LDV was performed with the simulation of the electric displacement field. The LDV measurements were performed on the length–width plane. The amplitude of the simulated displacement fields in the Y and X directions were extracted from the model using the line average function in COMSOL on the longer edge representing the length–width face of the PWAS. A good correlation between the modelled and measured spectra of the electromechanical displacement field is shown in Figure 12. Three different modes of vibration can be seen with distinct resonance frequencies around 10, 34 and 65 kHz in both the measured and modelled responses. The amplitude of the measured displacement fields was lower than that of the simulated response by a factor of 100. This was associated with friction, dielectric dissipation and losses, which were not considered in the model. A shift in the simulated frequency response was depicted, which increased almost linearly with the increase in frequency. This is possible because of the added mass from electrodes and electrical connections, which were not included in the model. Two other weaker modes were observed at 20 and 70 kHz in the measured response. These were associated to the vibration along the width and were not predicted by the model because of the plain strain 2D approximation.

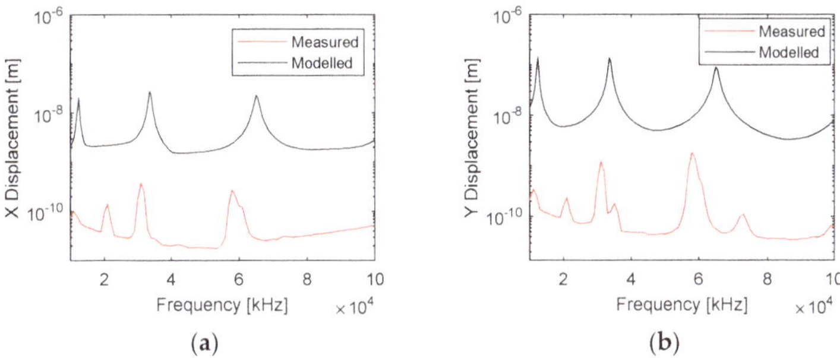

Figure 12. Comparison of the simulated displacement field with LDV measurements in a frequency range of 10–100 kHz showing displacement amplitudes in the (**a**) X and (**b**) Y directions.

Simulated velocity amplitudes from the length–thickness face were derived from the model and compared with LDV measurements. The modelled and measured vibration modes at 34, 65 and 100 kHz were compared and are shown in Figure 13. The modelled and measured results appeared to be in good agreement. The first two vibration modes at 34 and 65 kHz were longitudinal modes. The TS mode can be seen at 100 kHz, as the in-plane displacement was much larger than the out-of-plane displacement. In both the simulated and measured TS modes, a coupling of longitudinal mode can be observed. This is due to the boundary effects of the wrap-around electrode configuration. Acknowledging the knowledge of mode coupling and undesired miscellaneous modes, the model can be used to further study efficient backing materials or alternative electrode configurations to suppress them.

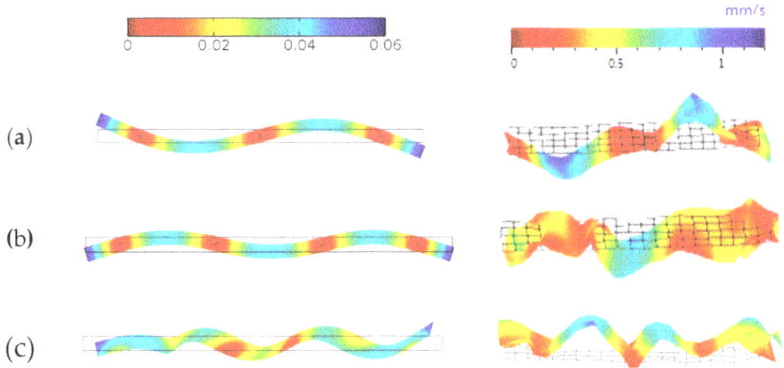

Figure 13. Vibration patterns of the Y-cut GaPO$_4$ PWAS simulated (left) and measured (right) for 34 kHz (**a**), 65 kHz (**b**) and 100 kHz (**c**). Colors indicate velocity amplitudes.

5. Discussions

This research focused on the examination of GaPO$_4$ for HT applications of UGWs by developing a TS-mode PWAS. An FEA model was developed and used to predict EMI and vibrational response accurately. The thermal response of the simulated EMI was validated with HT experiments and showed good agreement in terms of change in frequency and electromechanical coupling factor, which is a key FOM for ultrasonic applications. The model requires HT coefficients of material properties to simulate piezoelectric losses. If they are not available in the literature, a method for computing these coefficients for the TS mode was presented, which showed a good correlation with properties reported in the literature. To further improve the accuracy of the model, losses such as friction, damping and dielectric dissipation can be extracted from EMI response [42] and added to the model. The 2D model described here can be used to evaluate crystal cuts with no rotation. For single or multiple rotated crystal cuts, either 3D models with rotated coordinate systems or transformation of material properties will be required.

In UGW testing, signal interpretation can be very challenging because of the large number of wave modes that exist in a waveguide. Knowledge of vibration patterns of a transducer and how they are influenced by increasing temperature is crucial for transducer design. The model presented here predicted mode coupling and unwanted contour modes. This demonstrated the suitability of such a technique for optimising transducer design by suppressing unwanted modes.

The model can be further evaluated with time-domain analysis to analyse guided wave excitation on pipelines at HT. This study will require temperature-dependent properties of adhesives used for bonding. This will be useful for analysing excitation of any unwanted mode or transducer failure due to thermal stresses arising in a transducer as a result of differential thermal expansion.

6. Conclusions

In this study, $GaPO_4$ has been evaluated at temperatures up to 600 °C for UGW applications to superheated steam lines. $GaPO_4$ was fabricated into PWAS with dimensions, cut and electrode configuration to obtain the desired TS mode of vibration. The dielectric, elastic and piezoelectric properties corresponding to the TS mode were measured as a function of temperature to derive their temperature coefficients. An FEA model was developed to simulate the vibrational response and the temperature behaviour of the PWAS. These simulation results were validated experimentally with LDV and EMI measurements, respectively, where good agreement with the measured TS-mode shapes and the thermal response was observed. Finally, thermal aging experiments were conducted on the PWAS for over 1000 hours, and a stable TS-mode response was achieved, demonstrating its potential for SHM applications on steam lines. These results are of theoretical importance for the design of HT transducers for UGW applications.

Author Contributions: Conceptualization, A.D.; data curation, A.D.; funding acquisition, T.-H.G.; investigation, A.D.; methodology, A.D. and S.A.T.; resources, J.K. and T.-H.G.; supervision W.B. and T.-H.G.; validation, S.A.T. and A.D.; visualisation, A.D.; writing of the original draft, A.D.; writing of review and editing, W.B., J.K. and T.-H.G.

Funding: This research was partly funded by InnovateUK through the program for developing the civil nuclear power supply chain, as part of the UltraSteamLine project with the collaboration between Brunel University London and Plant Integrity Ltd.

Conflicts of Interest: The authors declare no conflicts of interest.

References

1.   Giurgiutiu, V. *Structural Health Monitoring with Piezoelectric Wafer Active Sensors*, 2nd ed.; Academic Press: Oxford, UK, 2014; ISBN 978-0-12-418691-0.
2.   Yu, L.; Santoni-Bottai, G.; Xu, B.; Liu, W.; Giurgiutiu, V. Piezoelectric wafer active sensors for in situ ultrasonic-guided wave SHM. *Fatigue Fract. Eng. Mater. Struct.* **2008**, *31*, 611–628. [CrossRef]
3.   Baptista, F.; Budoya, D.; Almeida, V.; Ulson, J. An Experimental Study on the Effect of Temperature on Piezoelectric Sensors for Impedance-Based Structural Health Monitoring. *Sensors* **2014**, *14*, 1208–1227. [CrossRef] [PubMed]
4.   Mei, H.; Haider, M.; Joseph, R.; Migot, A.; Giurgiutiu, V. Recent Advances in Piezoelectric Wafer Active Sensors for Structural Health Monitoring Applications. *Sensors* **2019**, *19*, 383. [CrossRef] [PubMed]
5.   Jiang, X.; Kim, K.; Zhang, S.; Johnson, J.; Salazar, G. High-temperature piezoelectric sensing. *Sensors* **2014**, *14*, 144–169. [CrossRef]

6. Tsiklauri, G.; Talbert, R.; Schmitt, B.; Filippov, G.; Bogoyavlensky, R.; Grishanin, E. Supercritical steam cycle for nuclear power plant. *Nucl. Eng. Des.* **2005**, *235*, 1651–1664. [CrossRef]

7. Gaur, B.; Babakr, A. Piping failure in a superheated steam service—A case study. *Matér. Tech.* **2013**, *101*, 206. [CrossRef]

8. Sabouri, M.; Hoseiny, H.; Faridi, H.R. Corrosion failure of superheat steam pipes of an ammonia production plant. *Vacuum* **2015**, *121*, 75–80. [CrossRef]

9. Xu, S.; Meng, W.; Wang, C.; Sun, Z.; Zhang, Y. Failure analysis of TP304H tubes in the superheated steam section of a reformer furnace. *Eng. Fail. Anal.* **2017**, *79*, 762–772. [CrossRef]

10. Bond, L.J. Fitness Tests For Old Nuclear Reactors–IEEE Spectrum. Available online: https://spectrum.ieee.org/energy/nuclear/fitness-tests-for-old-nuclear-reactors (accessed on 30 November 2019).

11. Liu, Z.; Kleiner, Y. State-of-the-art review of technologies for pipe structural health monitoring. *IEEE Sens. J.* **2012**, *12*, 1987–1992. [CrossRef]

12. Alleyne, D.N.; Cawley, P. The excitation of Lamb waves in pipes using dry-coupled piezoelectric transducers. *J. NDE* **1996**, *15*, 11–20. [CrossRef]

13. Guilded Ultrasonics Ltd. gPIMS®. Available online: https://www.guided-ultrasonics.com/gpims/ (accessed on 30 November 2019).

14. ANSI/IEEE Std 176-1987. IEEE Standard on Piezoelectricity. 1987. Available online: http://blogs.cimav.edu.mx/luis.fuentes/data/files/Curso_Cristalograf%C3%ADa/piezo_ieee.pdf (accessed on 9 December 2019).

15. Meeker, T. Publication and proposed revision of ANSI/IEEE standard 176-1987. *IEEE Trans. Ultrason. Ferroelectr. Freq. Control* **1996**, *43*, 717–772.

16. Lee, H.; Zhang, S.; Bar-Cohen, Y.; Sherrit, S. High Temperature, High Power Piezoelectric Composite Transducers. *Sensors* **2014**, *14*, 14526–14552. [CrossRef] [PubMed]

17. Kažys, R.; Voleišis, A.; Voleišienė, B. High temperature ultrasonic transducers. *Ultrasound* **2008**, *63*, 7–17.

18. Zhang, S.; Yu, F. Piezoelectric materials for high temperature sensors. *J. Am. Ceram. Soc.* **2011**, *94*, 3153–3170. [CrossRef]

19. Shinekumar, K.; Dutta, S. High-Temperature Piezoelectrics with Large Piezoelectric Coefficients. *J. Electron. Mater.* **2014**, *44*, 613–622. [CrossRef]

20. Gotmare, S.W.; Leontsev, S.O.; Eitel, R.E. Thermal Degradation and Aging of High-Temperature Piezoelectric Ceramics. *J. Am. Ceram. Soc.* **2010**, *93*, 1965–1969. [CrossRef]

21. Yu, F.; Chen, F.; Hou, S.; Wang, H.; Wang, Y.; Tian, S.; Jiang, C.; Li, Y.; Cheng, X.; Zhao, X. High temperature piezoelectric single crystals: Recent developments. In Proceedings of the 2016 Symposium on Piezoelectricity, Acoustic Waves, and Device Applications (SPAWDA), Xi'an, China, 21–24 October 2016; pp. 1–7.

22. Worsch, P.M.; Krempl, P.W.; Wallnofer, W. GaPO$_4$ crystals for sensor applications. In Proceedings of the IEEE Sensors, Orlando, FL, USA, 12–14 June 2002; Volume 1, pp. 589–593.

23. Palmier, D.; Cohior, R.; Bonjour, C.; Martin, G.; Zarembovitch, A.; Bigler, E.; Philippot, E. New results on the thermal sensitivity of bulk and surface modes of gallium orthophosphate GaPO$_4$. In Proceedings of the 1995 IEEE International Ultrasonics Symposium, Seattle, WA, USA, 7–10 November 1995; Volume 1, pp. 605–610.

24. Hamidon, M.N.; Skarda, V.; White, N.M.; Krispel, F.; Krempl, P.; Binhack, M.; Buff, W. Fabrication of high temperature surface acoustic wave devices for sensor applications. *Sens. Actuators Phys.* **2005**, *123–124*, 403–407. [CrossRef]

25. Thanner, H.; Krempl, P.W.; Selic, R.; Wallnöfer, W.; Worsch, P.M. GaPO$_4$ High temperature crystal microbalance demonstration up to 720 °C. *J. Therm. Anal. Calorim.* **2003**, *71*, 53–59. [CrossRef]

26. Krispel, F.; Reiter, C.; Neubig, J.; Lenzenhuber, F.; Krempl, P.; Wallnofer, W.; Worsch, P. Properties and applications of singly rotated GaPO$_4$ resonators. In Proceedings of the IEEE International Frequency Control Symposium and PDA Exhibition Jointly with the 17th European Frequency and Time Forum, Tampa, FL, USA, 4–8 May 2003; pp. 668–673.

27. Krempl, P.; Reiter, C.; Wallnofer, W.; Neubig, J. Temperature sensors based on GaPO$_4$. In Proceedings of the 2002 IEEE Ultrasonics Symposium, Munich, Germany, 8–11 October 2002; Volume 1, pp. 949–952.

28. Kostan, M.; Mohimi, A.; Nageswaran, C.; Kappatos, V.; Cheng, L.; Gan, T.-H.; Wrobel, L.; Selcuk, C. Assessment of Material Properties of Gallium Orthophosphate Piezoelectric Elements for Development of Phased Array Probes for Continuous Operation at 580 °C. *IOP Conf. Ser. Mater. Sci. Eng.* **2016**, *108*, 012008. [CrossRef]

29. Standard, B. Piezoelectric Properties of Ceramic Materials and Components-Part 2: Methods of Measurement–Low Power. Available online: https://infostore.saiglobal.com/en-us/Standards/EN-50324-2-2002-348917_SAIG_CENELEC_CENELEC_797147/ (accessed on 10 December 2019).

30. Cao, W.; Zhu, S.; Jiang, B. Analysis of shear modes in a piezoelectric vibrator. *J. Appl. Phys.* **1998**, *83*, 6. [CrossRef]

31. Frankel, D.J.; Bernhardt, G.P.; Sturtevant, B.T.; Moonlight, T.; Pereira da Cunha, M.; Lad, R.J. Stable electrodes and ultrathin passivation coatings for high temperature sensors in harsh environments. In Proceedings of the 2008 IEEE Sensors, Lecce, Italy, 26–29 October 2008; pp. 82–85.

32. Richter, D.; Fritze, H. High-temperature stable electrodes for langasite based surface acoustic wave devices. In Proceedings of the Sensor+Test Conferences, Nürnberg, Germany, 7–9 July 2011; pp. 532–537.

33. Hamidon, M.N.; Mousavi, S.A.; Isa, M.M.; Ismail, A.; Mahdi, M.A. Finite Element Method on Mass Loading Effect for Gallium Phosphate Surface Acoustic Wave Resonators. In Proceedings of the World Congress on Engineering, London, UK, 1–3 July 2009; pp. 447–452.

34. Reiter, C. Material properties of GaPO$_4$ and their relevance for applications. *Ann. Chim. Sci. Matér.* **2001**, *26*, 91–94. [CrossRef]

35. Silva, L.B.; Santos, E.J. Modeling high-resolution down-hole pressure transducer to achieve semi-distributed measurement in oil and gas production wells. *J. Integr. Circuits Syst.* **2019**, *14*, 1–9. [CrossRef]

36. Mohimi, A.; Gan, T.H.; Balachandran, W. Development of high temperature ultrasonic guided wave transducer for continuous in service monitoring of steam lines using non-stoichiometric lithium niobate piezoelectric ceramic. *Sens. Actuators Phys.* **2014**, *216*, 432–442. [CrossRef]

37. Dhutti, A.; Tumin, S.; Gan, T.; Kanfoud, J.; Balachandran, W. Comparative study on the performance of high temperature piezoelectric materials for structural health monitoring using ultrasonic guided waves. In Proceedings of the 7th Asia-Pacific Workshop on Structural Health Monitoring, Hong Kong, China, 12–15 November 2018; pp. 12–15.

38. Teletest Focus+ | Eddyfi. Available online: https://eddyfi.com/en/product/focus (accessed on 14 November 2019).

39. Hoshyarmanesh, H.; Ghodsi, M.; Kim, M.; Cho, H.H.; Park, H.-H. Temperature Effects on Electromechanical Response of Deposited Piezoelectric Sensors Used in Structural Health Monitoring of Aerospace Structures. *Sensors* **2019**, *19*, 2805. [CrossRef]

40. Nosek, J.; Pustka, M. About the coupling factor of the gallium orthophosphate, (GaPO$_4$) and its influence to the resonance - frequency temperature dependence. In Proceedings of the IEEE International Frequency Control Symposium and PDA Exhibition Jointly with the 17th European Frequency and Time Forum, Tampa, FL, USA, 4–8 May 2003; pp. 674–678.

41. Zarka, A.; Capelle, B.; Detaint, J.; Palmier, D.; Philippot, E.; Zvereva, O.V. Studies of GaPO$_4$ crystals and resonators. In Proceedings of the Proceedings of 1996 IEEE International Frequency Control Symposium, Honolulu, HI, USA, 5–7 June 1996; pp. 66–71.

42. González, A.; García, Á.; Benavente-Peces, C.; Pardo, L. Revisiting the Characterization of the Losses in Piezoelectric Materials from Impedance Spectroscopy at Resonance. *Materials* **2016**, *9*, 72. [CrossRef]

Article

Dry Coupling of Ultrasonic Transducer Components for High Temperature Applications

Neelesh Bhadwal [1,*], Mina Torabi Milani [1], Thomas Coyle [2] and Anthony Sinclair [1]

[1] Department of Mechanical and Industrial Engineering, University of Toronto, Toronto, ON M5S 3G8, Canada; mtorabi@mie.utoronto.ca (M.T.M.); sinclair@mie.utoronto.ca (A.S.)

[2] Department of Materials Science and Engineering, University of Toronto, Toronto, ON M5S 3E4, Canada; coyle@ecf.utoronto.ca

* Correspondence: neeleshb@mie.utoronto.ca

Received: 31 October 2019; Accepted: 4 December 2019; Published: 6 December 2019

Abstract: The viability for dry coupling of piezoelectric ultrasonic transducer components was investigated, using a thin foil of annealed silver as a filler material/coupling agent at each component interface. Criteria used for room temperature evaluation were centered on signal-to-noise ratio (SNR) and echo bandwidth, for a Li-Nb based transducer operating in pulse-echo mode. A normal clamping stress of only 25 MPa, applied repeatedly over three loading cycles on a precisely-aligned transducer stack, was sufficient to yield backwall echoes with a SNR greater than 25 dB, and a 3 dB bandwidth of approximately 65%. This compares to a SNR of 32 dB and a 3 dB bandwidth of 65%, achievable when all transducer interfaces were coupled with ultrasonic gel. The respective roles of a soft filler material, alignment of transducer components, cyclic clamping, component roughness, and component flatness were evaluated in achieving this high efficiency dry coupling, with transducer clamping forces far lower than previously reported. Preliminary high temperature tests indicate that this coupling method is suitable for high temperature and achieves signal quality comparable to that at room temperature with ultrasonic gel.

Keywords: ultrasonic transducer; elevated temperature; dry coupling; harsh environment; lithium niobate

---

1. Introduction

Ultrasonic transducers are used to inspect engineering components for flaws so that remedial action can be taken to avoid component failure. In some cases, it is desirable that the transducers be permanently mounted at critical piping/pressure vessel sites for continuous on-line monitoring of pipe integrity or characteristics of fluids inside the pipe. The pipe or pressure vessel may be very hot, such that the transducer must withstand high temperatures (and perhaps thermal cycling) for extended periods. Potential industrial applications include electrical power plants and petrochemical processing installations. Such a monitoring system reduces the need for plant shutdown to inspect the integrity of components and gives early warning of anomalous condition of the fluids inside the pipe or pressure vessel.

High-temperature ultrasonic inspection techniques in current use include the use of delay lines in front of the transducer to isolate the transducer from the high temperature [1,2]. Water cooling of the delay line or the transducer can also be used to increase the service temperature of traditional ultrasonic transducers by keeping the temperature of the ultrasonic transducer well below the temperature of the component being tested [2]. Electromagnetic acoustic transducers (EMAT) have been used up to 900 °C with active cooling, or 450 °C without active cooling, for continuous inspection [3,4]. EMATs; however, do suffer from high power needs, have poor signal-to-noise ratio, and are very sensitive to perturbations in sensor lift-off [5].

The overall long-term objective of this research project is to develop a direct-contact high-temperature ultrasonic transducer (continuous operation at up to 800 °C, with occasional thermal cycling) for non-destructive testing and process monitoring in the petrochemical industry. Currently, direct-contact non-destructive testing with ultrasonic transducers can only be done up to 550 °C without the use of cooling systems and delay lines [2,6]. Such commercially available high-temperature ultrasonic transducers for continuous use can cost several thousand dollars and are prone to failure after less than two years of use [5]. These transducers may fail due to a mixture of the following:

- Thermal stresses, causing cracking of components and breakdown of acoustic coupling between transducer components;
- Failure of individual components due to phase transitions or melting;
- Depoling of the piezoelectric element in the transducer by exceeding its Curie temperature.

Our previous studies focused on the selection and manufacturing of the backing, piezoelectric element, and matching layer for a cylindrically-shaped high-temperature ultrasonic transducer [5,7]. The suitable transducer components identified for high-temperature application were a rhombohedral 36° Y-cut lithium niobate piezoelectric crystal, a porous zirconia mechanical backing, and a stainless-steel 321 protective layer [5,7]. Methods for acoustically coupling the backing to the piezoelectric element were investigated in those earlier studies; a brazing system based on a silver-copper 72-28 braze (AgCu) was developed, but led to high reject rates of transducers due to cracking of the piezoelectric element [7,8].

The current study focuses on acoustically coupling the three principal transducer components together via dry coupling at each interface. Each pair of contacting materials must have similar acoustic impedance values for effective energy transmission. Air has an acoustic impedance value that is significantly lower than that of solid materials, such that any interfacial air gaps will block virtually all transmission of ultrasound from one transducer component to the next. Figure 1 shows the different components of the transducer stack and labels the three interfaces that were to be coupled. Interface (1)—backing–piezoelectric; interface (2)—piezoelectric–matching layer; and interface (3)—matching layer–test piece.

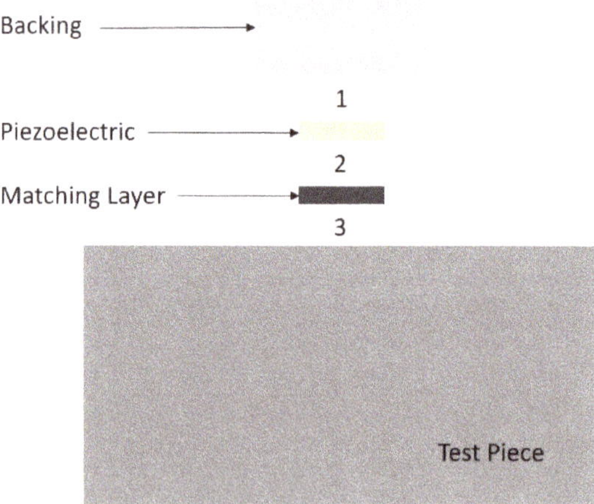

Figure 1. Transducer stack and interfaces.

For dry ultrasonic coupling of any two components, high clamping pressure is applied to the two contacting surfaces; the goal is to cause one or both components to slightly deform at the interface, in order to mate with the other component and expel any air trapped between them. It is critical that these high clamping loads be distributed uniformly over the entire interface, to avoid the prospect of good ultrasonically coupling over only a limited portion of the nominal interfacial area. Nonuniformly distributed loads also create bending moments in the contacting parts, which may lead to failure in brittle components such as the piezoelectric element.

For the cylindrical components in our transducer, the following distinct factors are important to achieve uniform load distribution across the interfaces linking ultrasonic transducer components [9]:

- Specimen ends must be flat, parallel, and perpendicular to the lateral surfaces.
- Diameters of the individual components should be constant throughout the component.
- Components should be concentrically aligned.

Once uniform load distribution is obtained, the magnitude of dry coupling pressure across an interface required for efficient ultrasonic coupling is dependent primarily on (1) surface finish (roughness), (2) plastic deformation of asperities on each surface, and (3) material hardness [10,11].

Surface finish: Rough surfaces with small narrow bumps will require sufficient clamping forces to flatten out the peaks and valleys at the interface in order to expel any air gaps (i.e., there must be localized yielding and small-scale plastic flow of the protruding material). The required clamping force to achieve good coupling is; therefore, expected to be higher if one or both contacting surfaces are very rough. Although the average dry coupling pressure is the clamping force applied to the transducer components divided by the cross-sectional area of the components, the pressure at individual protruding locations can be much higher. Therefore, the pressure on the protrusions can surpass the material yield stress, even at relatively low transducer clamping loads.

Plastic deformation of asperities: During the compression loading of an interface, plastic deformation of the asperities takes places at loads that correspond to nominal pressures significantly below the yield strength of the material [10]. When the load is subsequently reduced, the plastic deformation remains such that the two surfaces are now better "matched" to each other. If the interface is then loaded again to the same level as before, primarily elastic deformation of the asperities takes place [10,12]. Lower loads are; therefore, required during the second or third loading cycle for the same degree of acoustic transmission through an interface.

Material Hardness: An interface between relatively soft materials require a lower clamping pressure for dry coupling pressure, as the softer materials deform more easily. For an interface between hard materials, a soft "filler" material, such as a gold or silver foil, can be placed between the contacting surfaces to bridge any gap caused by the roughness or waviness of either surface [13]. The clamping pressure required for dry coupling is, thereby, reduced.

Several researchers suggest that the dry coupling pressure required for acoustic coupling of two metallic surfaces is in the order of a few hundred megapascals [10,12,14–16]. Such high stresses can damage the piezoelectric element as well as permanently deform the test piece, if the stress linking transducer components is also placed on the pipe or pressure vessel under inspection; at this point ultrasound inspection is no longer non-destructive.

For these reasons, techniques to minimize the required pressure for dry coupling were investigated. The objective of this study was to determine the dry coupling pressures required for each interface of the transducer stack. The dry coupling pressure was to be reduced from expected values in literature using proper alignment, cyclic loading, and soft filler materials, so as to make it a feasible coupling method in high temperature transducers. For this initial investigation, tests and analysis are conducted at room and elevated temperature, such that dry coupling performance can be compared to that achievable with ultrasonic gel couplant.

2. Materials and Methods

The components of the high-temperature ultrasonic transducer design were:

- Rhombohedral 36° Y-cut lithium niobate piezoelectric crystal: The discs were designed to resonate at 3 MHz and were obtained from Boston Piezo Optics Inc. in Bellingham, MA, USA. The surface of the LiNbO3 was fine-lapped and bare of any electrode material. The crystal had a diameter of 1 cm and a thickness of 1 mm.
- Porous zirconia mechanical backing: A novel process to create porous ceramic backings was reported in previous work, in which polyethylene beads of diameter 75–90 microns were mixed into yttria stabilized powder [5,7,17]. The powder was then mixed and pressed in a die to form a "green" body. During the subsequent sintering process, the beads vaporized and left behind spherical cavities (pores). The backing was 1 cm thick with a diameter of 15 mm.
- Stainless-steel 321 protective layer: Square matching layers (1.5 × 1.5 cm) were cut from shim stock of thickness 0.41 mm [7].

A block of low carbon steel was used as the test piece.

2.1. Alignment

In order to determine the dry coupling pressure required for each interface of the ultrasonic transducer, the entire stack (backing, piezoelectric element, protective layer) was subjected to compression loading in a load frame at room temperature. ASTM E9: Standard Test Methods of Compression Testing of Metallic Materials at Room Temperature states the following requirements for samples undergoing compression tests [9]:

1. If possible, cylindrical test specimens should be used.
2. Specimen surfaces should have a surface roughness of 1.6 µm (63 µin) R_a or better.
3. Specimen ends must be flat and parallel 0.0005 mm/mm (in./in.) and perpendicular to the lateral surfaces to within 3′ of arc.
4. Diameter of the specimen should not vary more than 1% of the specimen length or 0.05 mm (0.002 in).
5. The centerlines of all lateral surfaces of the specimens shall be coaxial within 0.25 mm (0.01 in).

Although this standard is meant for metals and is not directly applicable to our ultrasonic transducer, it does indicate the key parameters for compression tests. The transducer components for our study had a deviation from parallelism of less than 0.002 mm/mm.

To qualitatively assess whether the pressure distribution on the lithium niobate crystal was uniform, the transducer stack (consisting of backing, crystal, matching layer) was placed on top of the low carbon steel test piece in an Instron 3367 load frame with a 30 kN load cell. A piece of Fuji Prescale® Medium Film (obtained from Fujifilm Canada) was placed between the crystal surface matching layer. This film turns various shades of pink when pressures between 9.8 to 49.0 MPa (1400–7100 psi) are applied to it, according to charts provided by the manufacturer. The transducer stack was compressed, and a circle with a single shade of pink was formed on the film (Figure 2). Thus, a uniform load distribution was obtained, to within the manufacturer-specified uncertainty of ±10% [18].

Figure 2. Uniform shade of pink on pressure film.

2.2. Setup of Dry Coupling Tests

Further load frame tests were conducted to determine the dry coupling pressure required at each transducer interface if a filler material is used to achieve good signal transmission (Figure 3).

Ultrasonic gel was inserted at two of the interfaces, while a filler material foil was placed in between the two components whose dry coupling pressure was to be determined. Surface characteristics of the transducer components are listed in Table 1, where R_a is the average roughness and W_a is the average waviness (W_a) of the surface.

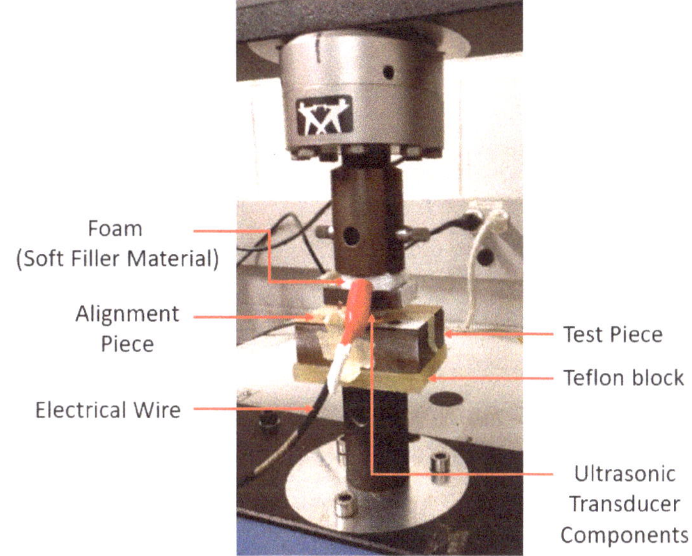

Figure 3. Load frame setup.

Table 1. Surface properties.

Component	Roughness (R_a) nm	Waviness (W_a) nm
Lithium Niobate	174	7
Matching Layer	198	1356
Test Piece (100 grit sandpaper)	798	1255

An Olympus 5077PR Square Wave Pulser/Receiver was used to excite the piezoelectric crystal with rectangular pulses of 300 V and a width of 17 µs. The pulses were sent into the test piece and an Agilent InfiniiVision DSO-X 2022A oscilloscope was then used to capture the backwall echo at each load increment of the load frame, for three loading cycles. A sampling frequency of 100 MHz was used, averaging 64 acquisitions per capture. MATLAB was then used to generate a normalized frequency spectrum using a Hann window with a length of 2 to 4 µs. It is important to note that the test piece was acoustically homogenous and thus no extra signal reconstruction was needed, as would be in the case of an acoustically heterogeneous material with attenuation distributions [19,20].

As shown in [10,12], repeated loading of the dry coupled ultrasonic transducer reduces the load required to obtain a specified level of acoustic transmission. As there was only a small change in acoustic transmission after the second loading cycle, it was decided that our ultrasonic transducer would be subjected to three loading cycles to reach the limiting case of plastic deformation of filler material and signal transmission across interfaces.

A filler material to be placed between the two components at each interface of the transducer was selected to meet the following criteria:

- Must be softer than components being coupled together;
- Chemically and physically stable in the presence of the transducer components;

- Must be an electrical conductor to act as electrodes for the lithium niobate crystal.

Transducers were mounted onto the test piece using ultrasonic gel; an investigation of transducer mounting techniques for various industrial applications is beyond the scope of this investigation.

Leading candidates for the filler material were gold and silver foil. Gold foil was more expensive and has an acoustic impedance of 62.6 MRayl, which was significantly higher than the transducer components (backing—20.7 MRayl [7], crystal—34.1 MRayl [21], matching layer—39.3 [22]) [23]. It was not expected that the filler material's bulk properties (such as attenuation) would have a significant impact on the transmission of ultrasound through the filler, as the thickness of the material (50 μm) was significantly less than that of the ultrasonic wavelength.

Silver, which has an acoustic impedance of 37.8 MRayl, was chosen as the filler material [23]. Silver foil 50 μm thick was obtained and tested in both annealed and as-rolled form. Both types of silver foil had a purity greater than 99.95% [24,25]. As-rolled silver is roughly four times harder than annealed silver (Vickers Hardness) [26].

3. Results

3.1. Maximum Performance Test

As a reference point, "optimal coupling" was defined as the transducer performance when ultrasonic gel was placed between each interface of the stack; the backwall echo from the test piece was recorded for that reference configuration. The stack was loaded up to 24 kN, equivalent to a pressure of 305.7 MPa on the crystal. The echo SNR and bandwidth of the echo were largely independent of interfacial loading when gel coupling was used to link the transducer components:

- SNR ~32 dB
- Echo signal bandwidth ~65%

This optimal bandwidth figure was consistent with previous studies where a lithium niobate crystal was brazed onto the porous zirconia backing with AgCu braze foil [7,8].

3.2. As-Rolled Silver Foil Test

For interfacial pressure tests with as-rolled silver foil, transducer stacks were loaded in increments of 2 kN to a maximum load of 24 kN; the backwall echo signal was recorded at each load increment.

The best performance achievable using as-rolled silver foil as an interfacial couplant/filler material was an echo SNR of 25 dB and 60% bandwidth. The pressure needed on the entire stack was ~18 kN (equivalent to 229.3 MPa on the crystal). Table 2 shows the individual dry coupling loads required at each interface to achieve this target.

Table 2. Dry coupling loads (as-rolled silver).

Interface	Load (kN)	Pressure on Crystal (MPa)
(1) Backing–Crystal	18	229.3
(2) Crystal–Matching Layer	6	76.4

It is noted that all the crystals loaded up to 24 kN developed cracks that originated at the crystal periphery. Although these transducers were still functional, the cracks have negative implications for long term transducer performance, and no further tests were conducted with the as-rolled solver foil as a coupling agent. An alternative dry coupling system was required that involves a lower coupling load.

3.3. Annealed Silver Foil Test

Transducer loading trials were conducted using the experimental configuration described in Section 3.2, but using annealed silver as opposed to as-rolled silver foil at each transducer interface.

The amplitude of the ultrasonic echo was continuously monitored as the axial load was increased. A cycling loading sequence was used: Once the echo amplitude had stabilized with increasing pressure on the transducer stack, the load was reduced to 0.5 kN, and the stack was then reloaded multiple times. The backwall ultrasonic echo was recorded at every kilonewton of load.

No cracks in the crystals were formed during these compression tests using annealed silver foil, as the dry coupling loads and associate pressure required to reach a stable echo amplitude were much lower than those required in the previous section using as-rolled silver foil. Table 3 shows the load required at each interface to achieve stable ultrasonic coupling, in the first and third loading cycles.

Table 3. Dry coupling loads (annealed silver).

Interface	Load (kN) During 1st Loading Cycle	Pressure on Crystal (MPa)	Load (kN) During 3rd Loading Cycle	Pressure on Crystal (MPa)
(1) Backing–Crystal	6	76.4	1	12.7
(2) Crystal–Matching Layer	3	38.2	1	12.7

In comparison to the dry coupling loads required when as-rolled silver was used as the filler material, the loads required with annealed silver are ~2 to 3 times lower (see Table 2).

An instability was noted in the frequency spectra of the echoes for very low loading levels with the use of annealed foil. However, the echo shape and frequency spectrum stabilized once loaded to the point where good interfacial contact was achieved.

Annealed Silver Foil Test Piece

An investigation of transducer mounting techniques for various industrial applications is beyond the scope of this investigation, but could be an indicator for an area needing further investigation.

In the following experiments, annealed silver was placed inside all three interfaces of the transducer stack and the test piece. The low carbon steel test piece was hand polished with 100 grit sandpaper—the same surface preparation technique used by the project's industrial sponsor to prepare the pipe surfaces on which ultrasonic transducers are mounted. The axial compressive force exerted by the load frame was increased in increments; the SNR and peak-to-peak voltage of the backwall echo signal were recorded at each load increment. The test was stopped when the peak-to-peak voltage stabilized. The load was then reduced to 0.5 kN and reapplied.

Figures 4 and 5 show the SNR vs force graphs for two different sets of transducer components Sets 1 and 2. By the third loading cycle, a load of only 2 kN (25.5 MPa) was enough to produce an echo that had stabilized with a maximum SNR near 65% in both samples, with a spectrum centered at 3 MHz.

It is noted that the load required to achieve stable coupling during the first loading cycle was different for Set 1 and Set 2. Additionally, the SNR achieved in Set 2 was larger than that obtained in Set 1. This may be due to the roughness of test piece 1 being pressed down and flattened after loading with the first transducer—a phenomenon reported in literature [10,12]. This effect is prominent for test pieces with larger air gaps in the roughness of the surface.

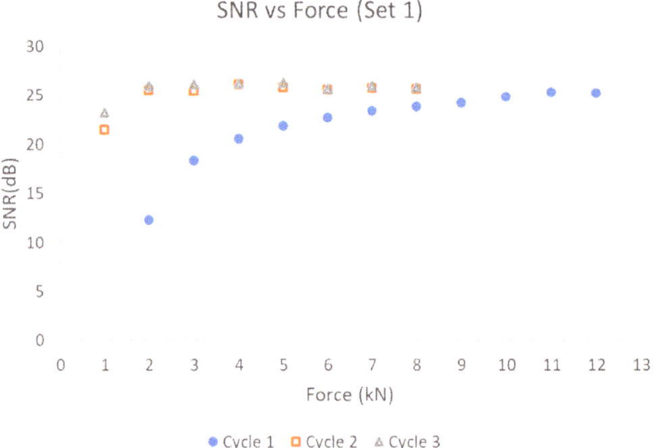

Figure 4. Axial loading for Set 1.

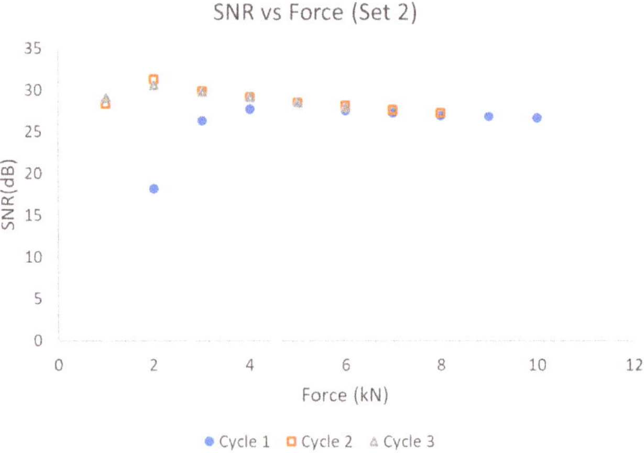

Figure 5. Axial loading for Set 2.

In summary, a rough, carbon steel surface typical of that found in an industrial plant can be acoustically coupled to our transducer prototype after cyclic loading three times with a load of 2 kN. When practical, pre-pressing of the surface can "flatten" surface roughness to a certain degree, making efficient dry coupling with annealed silver foil achievable at even lower loads.

3.4. High Temperature Coupling Test

A high temperature test was carried out to determine the suitability of this dry coupling method. Figure 6 shows a transducer assembly—a backing, piezoelectric crystal, and matching layer—coupled to a ground low carbon steel test piece. Annealed silver foil was placed in between the three interfaces (see Figure 1). The fixture was placed in the load frame and loaded up to 12 kN (152.9 MPa), unloaded, and then loaded to 8 kN (101.9 MPa). It was then unloaded and four screws with disc springs were tightened to a load of 3 kN (38.2 MPa). The disc springs were used to compensate for thermal expansion during the high temperature test. Spacers were machined to an exact height, so that when the screws were tightened, the washer on each bolt would touch the spacer when the springs provided a combined force of 3 kN on the transducer stack.

Figure 6. High temperature test fixture.

The fixture was placed in a furnace and heated up to 800 °C at 4 °C/min. The echo from the test piece was recorded at room temperature and every 100 °C. The SNR and bandwidth of the echo at each temperature increment are plotted in Figures 7 and 8.

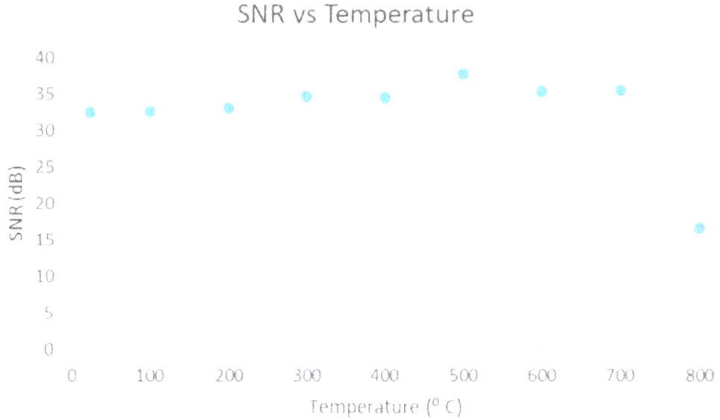

Figure 7. SNR vs temperature.

The SNR of the echo started at 32 dB and increased until 700 °C, after which there was a large drop. The bandwidth of the echo remained between 60% to 70% until 700 °C, after which there was a large drop. When the fixture was cooled and removed from the furnace it was noted that the screws were loose. The disc springs may have deteriorated at higher temperatures (their service temperature was 600 °C) and this could have led to a loss in dry coupling pressure resulting in the drop in signal quality at 800 °C. It is important to note that after the high temperature test was conducted the annealed silver foil was inspected and no signs of deterioration were observed.

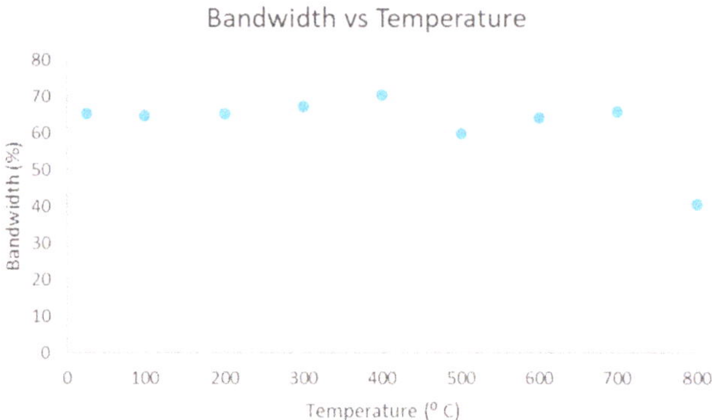

Figure 8. Bandwidth vs temperature.

To summarize the preliminary high temperature test suggests that this dry coupling method is suitable for high temperature applications, as the signal quality is comparable to that at room temperature with ultrasonic gel.

4. Discussion

The objective of this study was to determine the dry coupling pressures required for each interface of the transducer stack. Important criteria for proper transducer alignment were identified and a uniform load distribution was obtained when axially loading the transducer.

Two potential filler materials were identified to be placed between the component surfaces of the transducer stack as-rolled silver foil, and annealed silver foil.

With as-rolled silver foil, the dry coupling pressure required to produce an echo with an SNR of 25 dB and 60% bandwidth was 18 kN (229.3 MPa). Cracks were formed in the crystal during loading to 24 kN and thus further testing with cyclic loading was not carried out. Although these transducers were still functional, the cracks have negative implications for long term transducer performance and the dry coupling pressure was to be lowered.

The use of annealed silver foil reduced the dry coupling pressure required to couple the transducer layers together. The dry coupling pressure required between the backing and the crystal interface dropped by a factor of three and the dry coupling pressure required between the crystal and matching layer dropped by a factor of two. This is consistent with literature where drying coupling pressures depends on material hardness [11,13].

Cyclic loading of the transducer stack three times after initial assembly, with annealed foil as the filler material, reduced further the load required to 1 kN (12.7 MPa) in both interfaces. Cyclic loading caused the of surface deviations of the interfaces to be pressed down and flattened to a certain degree. Cyclic loading also caused the two surfaces to be better "matched" to each other, as shown in literature [10,12].

The few hundred megapascals dry coupling pressure suggested for acoustic coupling of two metallic interfaces can be greatly reduced through the use of precision aligned transducers, soft filler materials, and cyclic loading. Preliminary high temperature tests suggest that this method is suitable for high temperatures. Signals at high temperatures were comparable to that at room temperature. This may make dry coupling a feasible coupling method in high temperature transducers and may be an answer to a major problem in high temperature transducer manufacture.

For this initial investigation, tests and analysis are conducted at room and elevated temperatures, such that dry coupling performance can be compared to that achievable with ultrasonic gel couplant. Future work may be focused on the dry coupling of the transducer components with annealed silver or other filler materials with suitable properties at high temperatures. Future high temperature tests may include thermal cycling and filler material durability testing. Devising a scheme to keep constant pressure up to 800 °C will also be a target for future work.

Author Contributions: Conceptualization, N.B. and A.S.; methodology, N.B. and M.T.M.; formal analysis, N.B.; investigation, N.B.; resources, T.C.; writing—original draft preparation, N.B.; writing—review and editing, A.S.; supervision, A.S and T.C.; project administration, A.S.; funding acquisition, A.S. and T.C.

Funding: This research was funded by NSERC, Ontario Centres of Excellence's (OCE's) TalentEdge Internship Program (TIP) #29033 and Advanced Measurement and Analysis Group Incorporated.

Acknowledgments: Assistance by Advanced Measurement and Analysis Group Inc. in the form of technical support and experimental materials donated in kind were greatly appreciated.

Conflicts of Interest: The authors declare no conflicts of interest. The funders had no role in the design of the study; in the collection, analyses, or interpretation of data; in the writing of the manuscript, or in the decision to publish the results.

References

1. Emerson. Rosemount™ Wireless Permasense WT210 Corrosion and Erosion Monitoring System. Available online: https://www.emerson.com/en-ca/catalog/rosemount-sku-permasense-wt210-corrosion-erosion-monitoring-system (accessed on 4 December 2019).
2. High Temperature Ultrasonic Testing. Available online: https://www.olympus-ims.com/en/applications/high-temperature-ultrasonic-testing/ (accessed on 21 June 2019).
3. Lunn, N.; Dixon, S.; Potter, M. High temperature EMAT design for scanning or fixed point operation on magnetite coated steel. *NDT E Int.* **2017**, *89*, 74–80. [CrossRef]
4. Burrows, S.E.; Fan, Y.; Dixon, S. High temperature thickness measurements of stainless steel and low carbon steel using electromagnetic acoustic transducers. *NDT E Int.* **2014**, *68*, 73–77. [CrossRef]
5. Amini, M.H. Design and Manufacture of an Ultrasonic Transducer for Long-term High Temperature Operation. Ph.D. Thesis, University of Toronto, Toronto, Canada, November 2016.
6. Extreme Environment Permanent Monitoring Sensors & System Solutions. Available online: https://ionixadvancedtechnologies.co.uk/ (accessed on 21 November 2019).
7. Bosyj, C. Advancements in the Design and Fabrication of Ultrasound Transducers for Extreme Temperatures. Ph.D. Thesis, University of Toronto, Toronto, Canada, November 2017.
8. Bosyj, C.; Bhadwal, N.; Coyle, T.; Sinclair, A. Brazing Strategies for High Temperature Ultrasonic Transducers Based on LiNbO3 Piezoelectric Elements. *Instruments* **2019**, *3*, 2. [CrossRef]
9. American Society for Testing and Materials. *Standard Test Methods of Compression Testing of Metallic Materials at Room Temperature*; E9-19; ASTM International: West Conshohocken, PA, USA, 2019.
10. Drinkwater, B.; Dwyer-Joyce, R.; Cawley, P. A study of the interaction between ultrasound and a partially contacting solid—solid interface. *Philos. Trans. R. Soc. Lond. Ser. A* **1996**, *452*, 2613–2628.
11. Drinkwater, B.; Dwyer-Joyce, R.; Cawley, P. A study of the transmission of ultrasound across solid–rubber interfaces. *J. Acoust. Soc. Am.* **1997**, *101*, 970–981. [CrossRef]
12. Dwyer-Joyce, R.; Drinkwater, B.; Quinn, A. The use of ultrasound in the investigation of rough surface interfaces. *J. Trib.* **2000**, *123*, 8–16. [CrossRef]
13. Kažys, R.J.; Voleišis, A.; Voleišienė, B. High temperature ultrasonic transducers. *Ultragarsas* **2008**, *63*, 7–17.
14. Krolikowski, J.; Szczepek, J. Prediction of contact parameters using ultrasonic method. *Wear* **1991**, *148*, 181–195. [CrossRef]
15. Arakawa, T. A study on the transmission and reflection of an ultrasonic beam at machined surfaces pressed against each other. *Mater. Eval.* **1983**, *41*, 714–719.
16. Dwyer-Joyce, R. The application of ultrasonic NDT techniques in tribology. *Inst. Mech. Eng. Part J J. Eng. Tribol.* **2005**, *219*, 347–366. [CrossRef]
17. Amini, M.H.; Sinclair, A.N.; Coyle, T.W. Development of a high temperature transducer backing element with porous ceramics. In Proceedings of the 2014 IEEE International Ultrasonics Symposium, Chicago, IL, USA, 3–6 September 2014; pp. 967–970.
18. Specifications and Operational Environment. Available online: https://www.fujifilm.ca/products/measurement-films/prescale/film/#specifications (accessed on 23 May 2019).

19. Huang, C.; Nie, L.; Schoonover, R.W.; Wang, L.V.; Anastasio, M.A. Photoacoustic computed tomography correcting for heterogeneity and attenuation. *J. Biomed. Opt.* **2012**, *17*, 061211. [CrossRef] [PubMed]

20. Huang, C.; Wang, K.; Nie, L.; Wang, L.V.; Anastasio, M.A. Full-wave iterative image reconstruction in photoacoustic tomography with acoustically inhomogeneous media. *IEEE Trans. Med. Imaging* **2013**, *32*, 1097–1110. [CrossRef] [PubMed]

21. Material Properties Tables–UT–Piezoelectrics. Available online: https://www.nde-ed.org/GeneralResources/MaterialProperties/UT/ut_matlprop_piezoelectrics.htm (accessed on 12 May 2019).

22. MatWeb. 321 Stainless Steel, Annealed Sheet. Available online: http://www.matweb.com/search/datasheettext.aspx?matguid=303825a705054862bed3fe2dde947a5f (accessed on 2 August 2019).

23. Accoustic Properties for Metals in Solid Form. Available online: https://www.nde-ed.org/GeneralResources/MaterialProperties/UT/ut_matlprop_metals.htm (accessed on 12 May 2019).

24. Goodfellow. Silver–Foil (Thickness: 0.05mm Purity: 99.95+% Temper: As-rolled). Available online: http://www.goodfellow.com/catalogue/GFCat4I.php?ewd_token=6jqxOkVyyEJutto6GazISdgiM4WzY7&n=IecuSVeeCLGJu2wqitzi3o9IiMWFzD&ewd_urlNo=GFCat411&Catite=AG000300&CatSearNum=2 (accessed on 13 September 2019).

25. Goodfellow. Silver–Foil (Thickness: 0.05mm Purity: 99.95+% Temper: Annealed). Available online: http://www.goodfellow.com/catalogue/GFCat4I.php?ewd_token=6jqxOkVyyEJutto6GazISdgiM4WzY7&n=IecuSVeeCLGJu2wqitzi3o9IiMWFzD&ewd_urlNo=GFCat411&Catite=AG000305&CatSearNum=2 (accessed on 13 September 2019).

26. Goodfellow. Metal–Mechanical Properties. Available online: http://www.goodfellow.com/catalogue/GFCat2C.phpewd_token=p5dXczpXlb8bKNw8Jr5ELQuCaUZNRJ&n=ByIh4cJMPRbgwt4zEn6hKwYUbxxbRR&ewd_urlNo=GFCat26&type=00&prop=3 (accessed on 13 September 2019).

 *sensors*

Article

Design of a Phased Array EMAT for Inspection Applications in Liquid Sodium

Laura Pucci [1], Raphaële Raillon [1], Laura Taupin [1],* and François Baqué [2]

[1] French Alternative Energies and Nuclear Energy Commission—Laboratory for Integration of Systems and Technology (CEA-LIST) – Digiteo Labs, 91191 Gif-sur-Yvette Cedex, France; laura.pucci@cea.fr (L.P.); raphaele.raillon@cea.fr (R.R.)

[2] French Alternative Energies and Nuclear Energy Commission—Division of Nuclear Energy (CEA-DEN) – 13108 Saint-Paul-lez-Durance cedex, France; francois.baque@cea.fr

* Correspondence: laura.taupin@cea.fr

Received: 27 August 2019; Accepted: 9 October 2019; Published: 15 October 2019

Abstract: This article describes the development of a French CEA in-house phased array Electro Magnetic Acoustic Transducer (EMAT) adapted to hot and opaque sodium environment for in-service inspection of Sodium Fast Reactors. The work presented herein aimed at improving in-service inspection techniques for the ASTRID reactor project. The design process of the phased array EMAT is explained and followed by a review of laboratory experimental test results.

Keywords: EMAT sensor; Phased Array; L-waves; liquid sodium; inspection; NDE; SFR

---

1. Introduction

In-service inspection of Sodium Fast Reactors (SFR) requires the development of non-destructive techniques adapted to the hot (about 200°C) and opaque sodium environment. Piezoelectric probes are usually used for in-service inspection; however, in liquid sodium, this kind of probes encounters wetting problems. The liquid sodium does not adhere to the surface of the probe, leading to the presence of gas at the probe's surface that impedes an effective transmission of the ultrasounds in the liquid sodium. To deal with this issue, piezoelectric probes have to be immerged several hours in liquid sodium at very high temperature, which leads to damaging risks for these probes. Electro Magnetic Acoustic Transducers (EMAT) have shown promise for overcoming these wetting problems, as their technology enables the generation of pressure waves directly inside the liquid sodium, and therefore, using EMATs simplifies and accelerates the implementation of inspections in such environment. The R&D work concerning non-destructive inspection techniques for the ASTRID reactor (see [1] for an overview of the current status of the associated R&D program) has therefore aimed at exploring the potential of the EMAT technique, especially concerning under-sodium viewing and in situ distance measurements in the reactor pressure vessel. In that context, other ultrasonic transducers, based on piezoelectric concept, are being developed for Non-Destructive Examination (TUCSS of FRAMATOME [2]) and for Under Sodium Vision (IMARSOD of CEA [3]).

Previous work performed at the CEA-LIST Non-Destructive Testing Department has led to proofs of concept and liquid sodium feasibility tests using single element EMATs, as well as phased array EMATs [4–6]. These probes happened not to be sensitive enough, and the recent work has been focused on the ways to overcome this lack of sensitivity. This eventually led to a 12-element phased array EMAT sensitive enough to consider trials in liquid sodium.

In this paper, the development process of a 12-element optimized phased array EMAT will be first described, then, testing results obtained in the laboratory on aluminum blocks with this probe will be presented.

2. Context

Since France in 2008 considered the SFR concept to be the most mature for Generation IV nuclear reactors, an extensive R&D program was launched. In-Service Inspection was identified as a difficult task to perform (as sodium coolant is opaque, hot, and highly chemically reactive, as well as being difficult to drain). Ultrasonic techniques have been extensively studied, as they are well adapted to Non-Destructive Examination and telemetry measurement in this harsh environment.

Thus, development of ultrasonic transducers to be immersed in sodium at about 200 °C in the reactor block (inspection is performed at shutdown conditions) led to a first phase of specification consolidation, followed by a pre-qualification process involving increasingly more realistic experiments using acoustic techniques and simulations performed with the patented CIVA code [7].

Associated applications for inspection deal with telemetry, vision and volumetric control for SFR reactor block systems, structures and components, and also for the power conversion system. EMAT concept appears as an efficient candidate for alternative technology of piezoelectric ultrasonic sensors.

Previous work has led to an 8-element phased array EMAT, tested under liquid sodium environment at 180 °C for at least 24 h [6]. The EMAT was located 160 mm from the target, as shown in Figure 1. The target presents two parts: the first one is composed of a flat plate, and the second of a 6-mm-diameter rod. The flat plate is used as a reference, as all the energy sent to this part comes back to the sensor if its surface is parallel to the plate. Thus, this reference is used to adjust the orientation of the probe relative to the target. Then, the second part of the target is used to verify the angular beam deflection of the EMAT and its resolution on a rod. An angular scan with several positions was performed. The obtained image is shown in Figure 1. An echo is observed after the plate echo; this comes from the container wall behind the target.

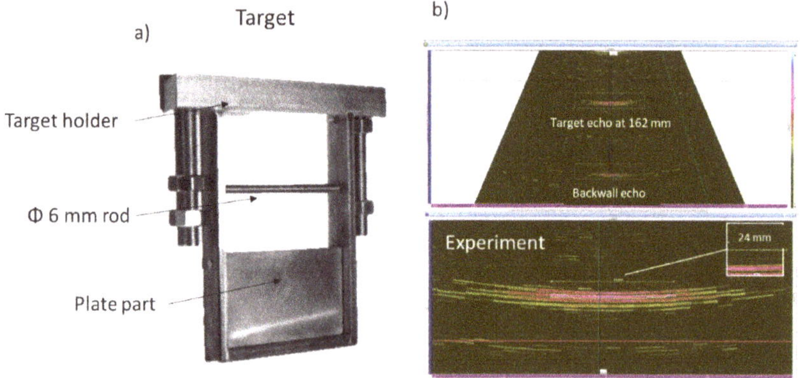

Figure 1. (**a**): Picture of the target used for liquid sodium test with the previous EMAT. (**b**): Ultrasonic scanning images during liquid sodium testing.

The obtained image suffers from a high noise level, which does not allow discrimination of the finer details of the advanced target (the 6 mm tube, for instance). The aim of the work described in this paper is to improve the sensitivity of the EMAT in order to be able to detect this kind of detail. However, these results with low sensitivity still demonstrate the proper functioning, over several hours, of the EMAT in a harsh environment (liquid sodium at 180 °C).

3. Development of the Optimized 12-Element Phased Array EMAT

The principle of EMAT force generation in the context of under sodium viewing is explained in Figure 2. A static magnetic field is induced in the region of the liquid sodium by permanent magnets. In liquid, only Longitudinal-waves (L-waves) can be generated, in which case the orientation of the static magnetic field delivered by permanent magnets is suitable for generating such ultrasonic waves. In addition to this field, eddy currents are generated directly inside the liquid sodium thanks to the coil, which is located between the two magnets, powered by an alternative current. The penetration depth of the eddy currents depends on the material conductivity and permeability and the pulse frequency of the coil alimentation. This is called skin depth. In liquid sodium, for a pulse frequency of 1 MHz, the skin depth is approximatively 110 µm. The cross product between the eddy current distribution and the static magnetic field results in a pulsed volume Lorentz force, which is located in the skin depth, and acts as a source for pressure waves. This force can be expressed by the following equation: $\vec{F}_{Lor} = q.\vec{v}\vec{B}$ (Figure 2). These ultrasonic pressure waves can then be used for inspection purposes in the liquid sodium. Potential applications include object detection, distance measurements, surface metrology, imaging for exploratory purposes and non-destructing testing of components. The design shown in Figure 2 supports the generation of ultrasonic vertical L-waves.

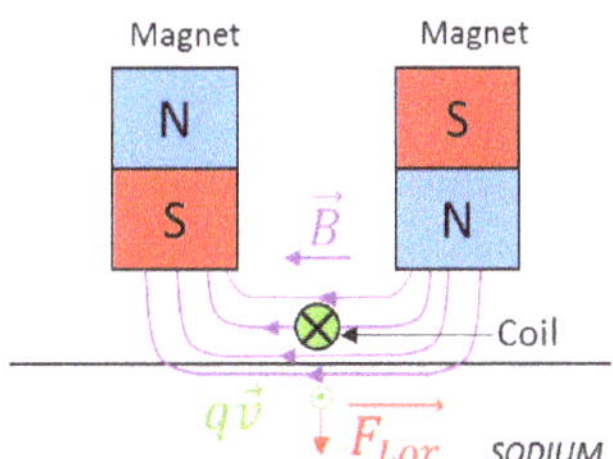

Figure 2. Physical principles of pressure wave generation in liquid sodium. $q\vec{v}$ is the electric current density, $\vec{B}$ the magnetic field, $\vec{F}_{Lor}$ the Lorentz force.

Based on this principle, a phased array EMAT was developed. The aim is to find the most suitable arrangement for the magnets and the coils for generating L-waves that are able to focus or deflect the L-beam and reach a sufficient sensitivity. This last point requires precisely considering the electronic aspect, which is essential for this kind of probe, the low transduction efficiency of which is well known.

3.1. Magnet and Coil Design

The phased array EMAT is made up of several coils. Each coil represents an element to which a delay can be applied in order to focalize or deflect the L ultrasonic beam by applying the appropriate delay laws to all the elements.

The challenge for this EMAT is to combine the need for a small pitch (the pitch is the distance between the centers of two adjacent elements (coils)) to ensure constructive interferences of the ultrasonic waves, and the need for large magnets and coils to ensure a sufficient sensitivity. Indeed, in order to be able to deflect the ultrasonic beam without increasing the grating lobes, it is recommended to fix the pitch "p" so that p/λ is lower than 1/2. With "λ" being the wavelength of the L-waves in liquid sodium that satisfies the formula λ = c/f, where "c" is the celerity of the L-waves in liquid sodium, and "f" is the center frequency of the EMAT excitation signal. For the targeted frequencies (1 to 2 MHz), this constraint imposes a pitch lower than 1.23 to 0.6 mm.

Two designs were considered for the phased array EMAT probe; they are represented in Figure 3.

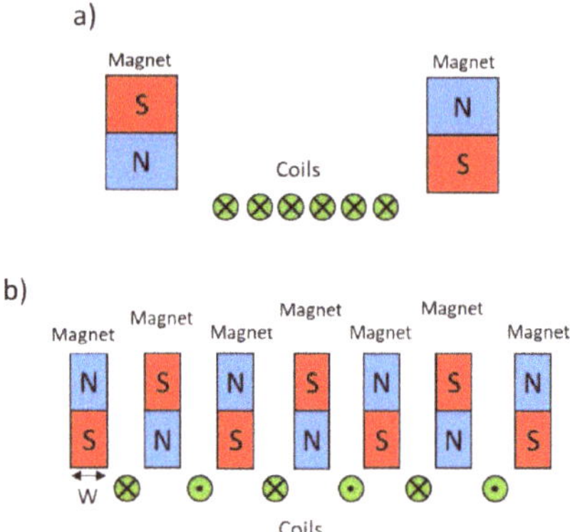

Figure 3. Diagram of two possible designs of the EMAT probe.

In diagram a), two strong magnets surround all of the elements (coils). This is the design of the first EMAT built for this project. Its main advantages are a small pitch and a strong magnetic field, which is achieved thanks to the large magnets. The main disadvantages are due to the large distance between the two magnets. This leads to an inhomogeneous magnetic field over the elements: the amplitude of the Lorentz forces generated by each coil decreases rapidly with distance from the magnet, preventing a good mastery of the beam. It also limits the number of elements. Because of these limitations, the other arrangement (b)) for the EMAT was chosen.

In diagram b), each element (coil) is surrounded by two magnets, allowing the same magnetic field over each element. Moreover, there is no more issue concerning the increase of the number of elements. However, the size of the magnets (width, "W" in Figure 3, is limited here, because the larger their size, the larger the pitch, and the greater the amplitude of the grating lobes. Furthermore, small magnets lead to weak magnetic fields, and therefore to weak probe sensitivity. Therefore, a compromise should be found between the pitch value and the magnets size. Please note that the same considerations remain valid for the coil width, as it impacts the pitch. Moreover, another consideration has to be taken into account regarding the coil width. Indeed, the EMAT is planned to operate in both emission and reception. In emission, a large coil width is needed to allow the coil to be crossed by high currents that will generate intensive ultrasonic waves, but in reception, a narrow coil width is needed to increase the sensitivity. These two antagonistic requirements impose a trade-off with respect to the coil width.

It can be noticed that two coils next to each other are supplied by current in opposite directions. This is due to the fact that the direction of the static magnetic field above one coil is in a direction opposite to that above the other. Then, in order to obtain constructive Lorentz forces, the current flowing through two adjacent coils needs to be in opposite directions.

Some simulations were performed with CIVA software in order to define the coil and magnet sizes for a 1 MHz excitation signal in liquid sodium. The results led to the following values: the magnets were 3 mm wide, 25 mm in height and 23 mm in length, and the coils width was 1.05 mm. The resulting pitch was 4.05 mm.

This value of the pitch, 4.05 mm, is significantly superior to the recommended value. The consequence is the presence of important grating lobes as displayed later.

3.2. Increase of the Magnetic Field

Once the dimensions of the coils and the magnets were chosen, the increase of the magnetic field was studied. Magnets magnetized along the $\vec{X}$ axis were interleaved between the magnets magnetized along the $\vec{Z}$ axis (Figure 4) in order to increase the magnetic field along the $\vec{X}$ axis at the surface of the sodium. This kind of design is called a "Halbach array" [8], and is depicted in Figure 4.

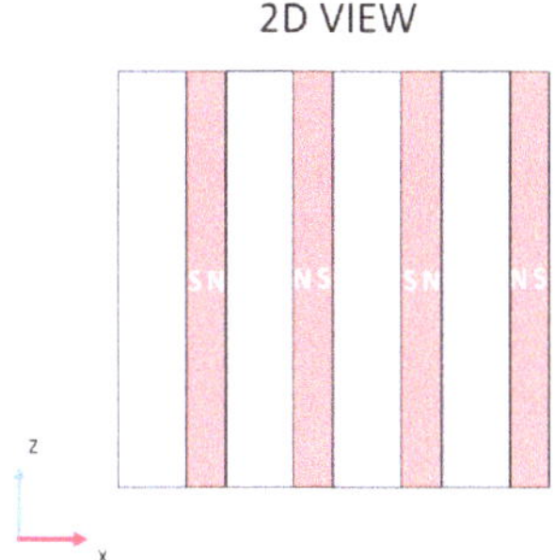

Figure 4. Magnetic assembly design of the EMAT probe with the "Klaus Halbach" design.

In Figure 5, diagrams of the EMAT are exposed. The twelve rectangular coils are etched on a flexible PCB and made of 8 turns. Each coil represents one element.

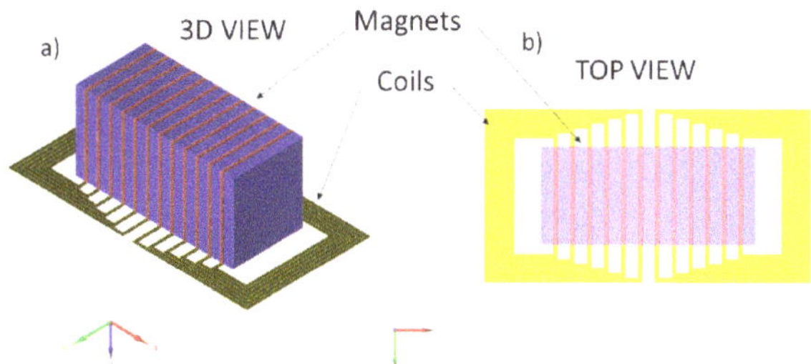

Figure 5. (**a**): 3D view of the EMAT. (**b**): top view of the EMAT.

This new magnet arrangement increases the magnetic field close to the surface (0.1 mm) by about a factor of 2 and by a factor of 1.5 at higher distances (1 mm).

The cartography of the amplitude of the vertical component of the Lorentz force generated below the 12-element EMAT was computed with CIVA in liquid sodium (Figure 6). As can be seen, the amplitude of the vertical components of the Lorentz force distribution are the same for each element, as required.

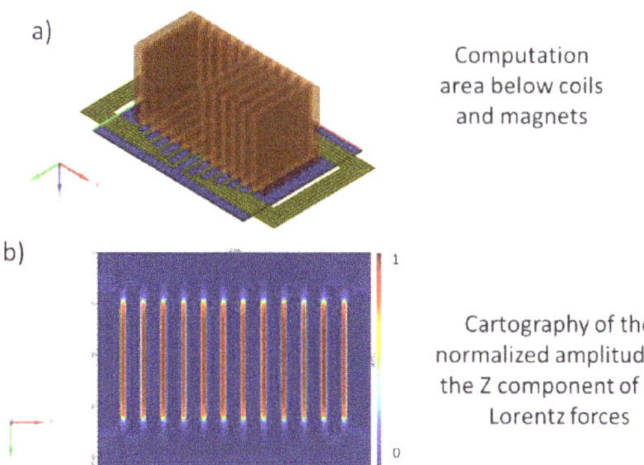

Figure 6. Simulation result. Normalized Z component of the Lorentz forces of the 12-element EMAT in liquid sodium. (**a**): Computation area below the sensor. (**b**): Cartography of normalized amplitude of the Z component of the Lorentz forces.

The magnets used in this study (laboratory experiments) were Neodymium-Iron-Boron magnets, because they are well known for delivering a strong magnetic field at room temperature. However, their magnetizing force decreases with temperature increase until they lose their magnetism at around 590 °C. Magnets with a higher Curie point should be used for measurements in liquid sodium at 180 °C. For the sodium trials, we used Samarium-Cobalt magnets with a Curie point at around 1000 °C.

3.3. Resonance Phenomena

After the choice of the geometric parameters of the EMAT, the electronic aspect was studied. The equivalent electronic model proposed by Jian et al. [9] for one coil is shown in Figure 7.

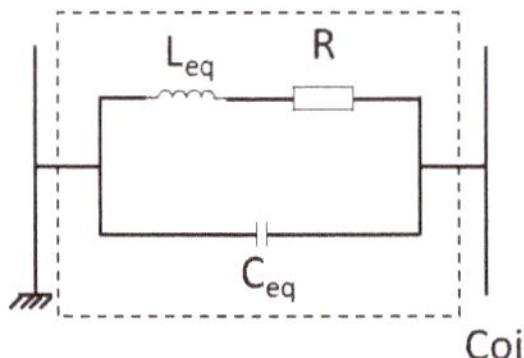

Figure 7. Equivalent electronic model proposed by Jian et al. [9] for one element of the EMAT.

This electronic circuit presents a resonance frequency described by Thomson formula, where L_{eq} is the equivalent inductance of the coil and C_{eq} is the equivalent capacity of the coil.

$$f = \frac{1}{\sqrt{2\pi L_{eq} C_{eq}}}$$

To improve the amplitude of the emitted and received signals of the EMAT, the resonance frequency formula must be verified for each coil at the center frequency "f" of the EMAT excitation signal. To adjust the parameters of the coil (L_{eq} and C_{eq}) to the work frequency (f) the value of the capacitor can be modified by adding a parallel capacitor, named C_{res}, as represented in Figure 8. The determination of the value of C_{res} is made theoretically for each coil by resolving the Thomson formula (knowing f and L_{eq}). The theoretical value is then adjusted experimentally by a current measurement in each coil.

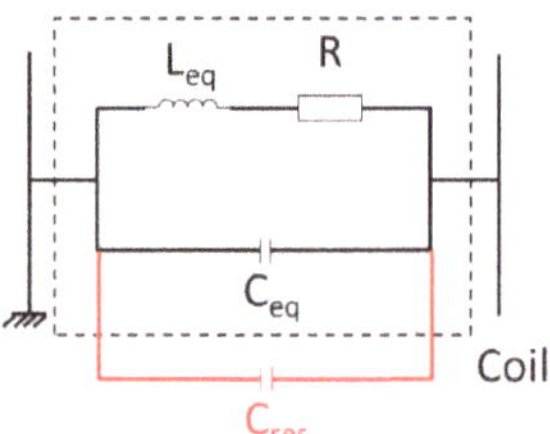

Figure 8. Equivalent model of a coil with the addition of a capacitor in order to operate at the resonance frequency.

In this second section, the geometry of the phase array EMAT and the method for optimizing the sensitivity of the probe (magnet arrangement and use of the resonance phenomena) have been described.

4. Characteristics of The Ultrasonic Beam in Liquid Sodium with Civa Simulation Software

In this third part, the acoustic beam characteristics of the EMAT were checked thanks to the CIVA software. The beam radiated by the 12-element phased array EMAT in liquid sodium at 180 °C was computed with CIVA. Different delay laws were applied to the EMAT in order to check its capacity to focus and deflect the beam.

4.1. Focusing without Angular Deflection

A first set of delay laws was applied in order to focus the L-waves of the ultrasonic beam at depths of 80, 100 and 150 mm without angular deflection (delay law called "L0°"). The results of the simulated beams are shown in Figure 9. The cartographies displayed on this figure show the maximum amplitude of the temporal acoustic signal radiated at each point of the cartography. The color blue corresponds to the maximum amplitude and the color green to the smallest amplitude.

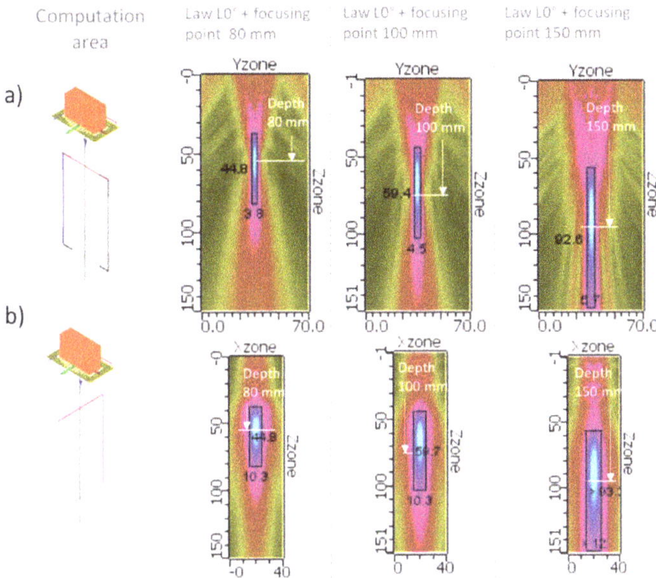

Figure 9. Simulation results. Longitudinal ultrasonic field radiated in liquid sodium by the 12-element EMAT. Delay laws applied to focus at 80 mm, 100 mm and 150 mm without beam deflection (law L0°). Computation of the field in two 2D areas located in the focal plane (figures a)) and in the orthogonal plane (figures b)). Excitation signal at 1 MHz composed of 3 bursts (signal close to the experimental one used in Section 4).

In the focal plane (images a) in Figure 9), the focal length varied from 45 mm to 93 mm when the focusing depth increased. The focal distances required in the 3 cases were reached. In the case of a focusing point at 150 mm, the presence of grating lobes very close to the main beam can be noticed. As it is not recommended to work with such an ultrasonic beam, a shorter focusing distance will be chosen.

In the orthogonal plane (images b) in Figure 9), the focal width varied from 10.3 to 12 mm when the focusing depth increased.

For each of those three simulations, the amplitude of the ultrasonic beam displayed is normalized. Nevertheless, a comparison of the three maximum radiated amplitudes was carried out. The simulation results show that the maximal amplitude of the beam decreased with the increase of the focal depth point. There was a loss of 3 dB between 80 mm and 150 mm focusing depths.

Considering the ultrasonic beam profile and amplitude, a focusing point at 100 mm was chosen for the next simulation step.

4.2. Focusing with Angular Deflection

A second set of delay laws was applied in order to focus and deflect the L-wave beam. The focal distance chosen was 100 mm and the deflection angles were 0°, 10° and 20°. The simulated beams are shown in Figure 10.

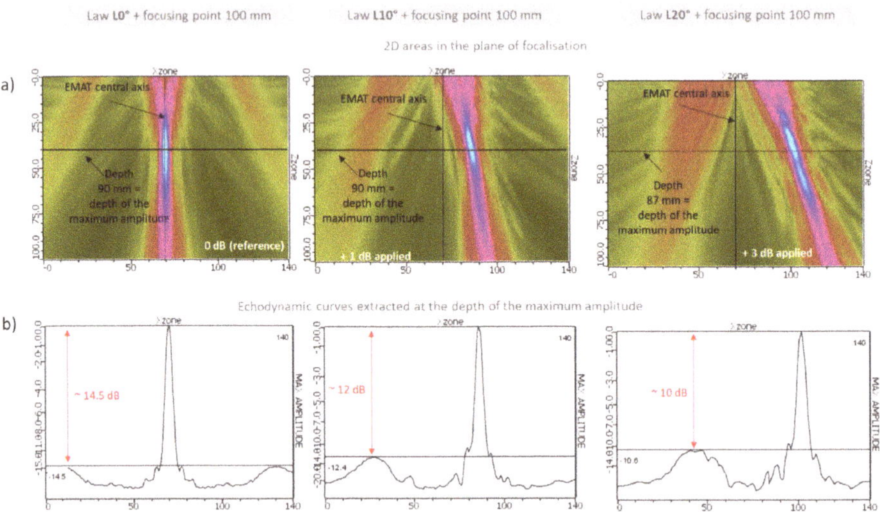

Figure 10. Simulation results. Longitudinal ultrasonic field radiated in 180 °C liquid sodium by the 12-element EMAT. Focusing law at 100 mm without deflection (left figure), with 10° deflection (middle figures) and with 20° deflection (right figure). Computation of the field in a 2D area located in the focal plane (figures a)) and amplitude of the field along a horizontal profile extracted from the 2D field ate the depth of the maximal amplitude (figures b)). Excitation signal at 1 MHz composed of 3 bursts (signal close to the experimental one used in Section 4).

These maps show that the beam was well deflected. However, as expected, because of the large value of the pitch with respect to the operating frequency, the amplitude of the grating lobes was significant and increased with the required deflection angle. Thus, it will be important to ensure that they do not lead to erroneous interpretation of the echoes, and simulations will be helpful for that.

The simulation results shown in Sections 2 and 3 validate the parameters chosen for the EMAT. However, CIVA does not predict EMAT sensitivity. The EMAT sensitivity was then evaluated on the basis of experimental trials performed on aluminum blocks.

5. Validation of Probe Performances Under Laboratory Conditions

In this section, the experimental results obtained with the EMAT on aluminum blocks are reported. The frequency used for these experimental measurements was 2 MHz, in order to obtain about the same wavelength in aluminum at this frequency (3.2 mm) as in liquid sodium at 1 MHz (2.5 mm).

5.1. Transmission Measurement on Aluminum Block

A first objective was to characterize the radiated beams obtained with different delay laws. Therefore, an 80-mm-thick aluminum block was used in a transmission configuration, with the EMAT and a fixed emitter located on one side of the block, and a piezoelectric probe and receiver scanning the surface on the other side of the block. This configuration is shown in Figure 11.

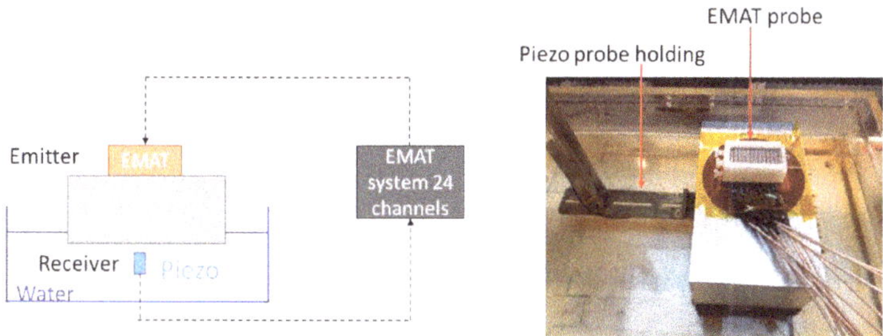

Figure 11. Experimental setup for transmission measurement—emission by the EMAT and reception by the piezoelectric probe.

Five delay laws were applied for focusing the longitudinal wave beam radiated at the backwall of the block (80 mm) with different angles of deflection ($-20°$, $-10°$, $0°$, $10°$ and $20°$). An alternative voltage of 250 V at 2 MHz composed of 3 bursts supplied the coils of the EMAT. The signals measured by the piezoelectric transducer placed close to the backwall of the aluminum block are presented in Figure 12. The C-scans displayed represent the maximum amplitude of the signal recorded with the piezoelectric probe at each position of the scan.

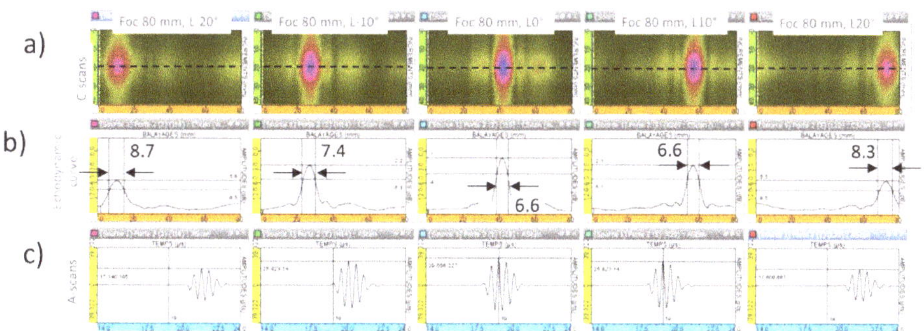

Figure 12. Experimental results obtained for five delay laws applied to focus the L beam at 80 mm depth with angular deflections of L-20°, L-10°, L0°, L10° and L20° (from left to right). (**a**): C-scans; (**b**): Echodynamic curves; and (**c**): A-scans extracted at the position of the maximum amplitude of the C-scans.

It can be noticed that the ultrasonic beams were deflected as expected, with a good spatial resolution. The focal width at -3 dB does not increase much with the deflection, as can be seen on the echodynamic curves (6.6 mm for L0° and 8.7 mm for L20°). The maximal amplitude radiated at 80 mm decreases with deflection: there is a loss of 4.6 dB between the L0° and the L20° deflections.

In addition, the amplitude of the grating lobes observed around the main beam are approximatively 12 dB weaker than the maximal amplitude at the focal point.

5.2. Pulse Echo Measurement on Aluminum Block

In this section, the results obtained with the EMAT probe in pulse echo mode are exposed. The aluminum block used is the same as the one described in Section 4.1. The alimentation of the coils also remained identical. This configuration is shown in Figure 13.

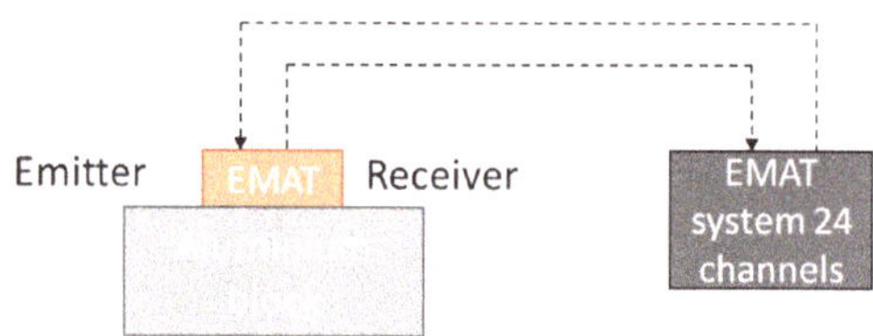

Figure 13. Experimental setup for pulse echo measurement—emission and reception by the EMAT.

The L-wave backwall echo of the block described in Section 4.1 and received when no delay law was applied in either emission or reception was measured. The A-scan obtained is presented in Figure 14.

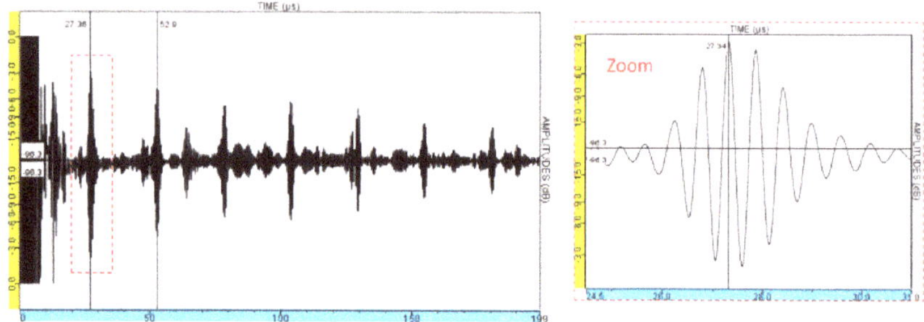

Figure 14. Experimental measurements: A-scan obtained with the EMAT probe in pulse echo mode. No focusing, aluminum block of 80 mm.

Considering the velocity of the longitudinal ultrasonic waves in aluminum, approximatively 6300 m/s, and the block thickness, 80 mm, the echo observed at 27 µs is the first backwall echo. The second backwall echo is observed at 52.9 µs. Several echoes of weaker amplitude that appear between the multiple backwall echoes can be noticed.

Then, the echoes of five Side-Drilled Holes (SDH) 6 mm in diameter located at 180 mm depth in a 250-mm-thick aluminum block were measured. The characteristics of this block and the experimental setup are shown in Figure 15.

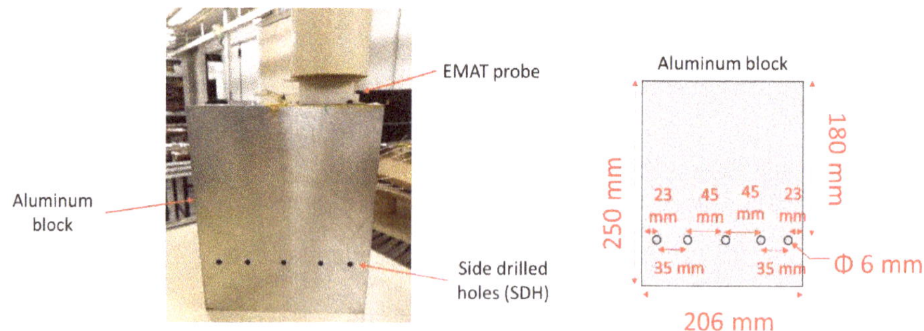

Figure 15. Experimental setup and geometric characteristics of the aluminum block used for the echo measurements on side-drilled holes.

Fifty delay laws were applied in order to realize an angular deflection of the L-wave beam (sectorial scanning) between −20° and +20° with a focusing at 180 mm. This distance corresponds to the depth of the SDH. It is greater than the near-field distance of the EMAT, but this block was the only aluminum block with SDH available for this study. This sectorial scanning configuration is presented in Figure 16.

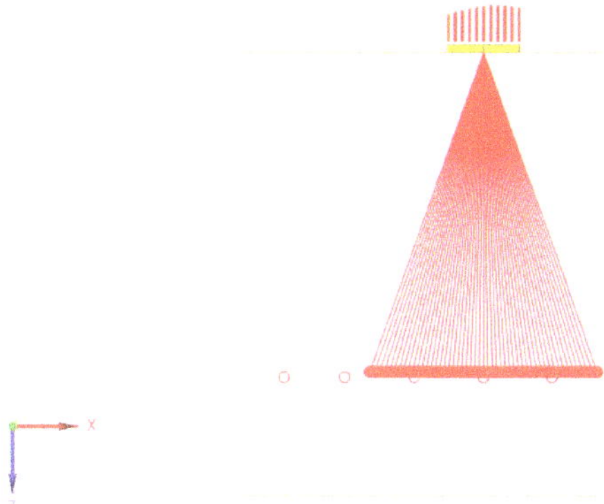

Figure 16. Representation of the sectorial scanning configuration for the inspection of the side-drilled holes.

In Figure 17, the measured S-scan (Sectorial scan) is represented: in the figure a), the S-scan is reconstructed in the block using the L-waves time of flight and on the figure b) the S-scan is displayed without reconstruction with the fifty laws applied on the x-axis and the time on the y-axis.

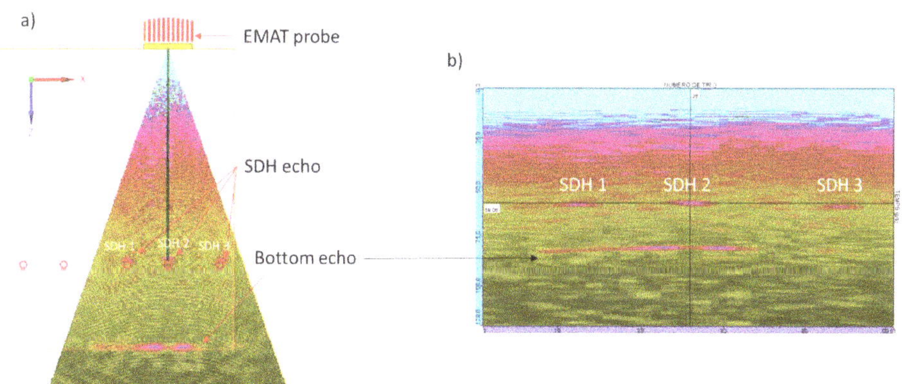

Figure 17. Experimental results. S-scans obtained on the aluminum block (250 mm thickness) with 3 SDHs (6 mm diameter) at 180 mm depth.

The A-scans extracted from the S-scan at the maximum amplitude of each SDH's echo are presented in Figure 18.

It may be observed that the dead zone spreads at over 40 µs. Beyond this zone, the 3 SDH responses were detected, as well as the response of the backwall of the block. In each A-san, both echoes are marked with a vertical cursor.

As expected, the echo of the SDH 2 (identification on Figures 17 and 18) is detected with the best SNR (Signal to Noise Ratio), as it is obtained using the non-deflected beam, whereas the SDH 1 and SDH 3 are detected, with deflections of 11.5° and 15.2°, respectively.

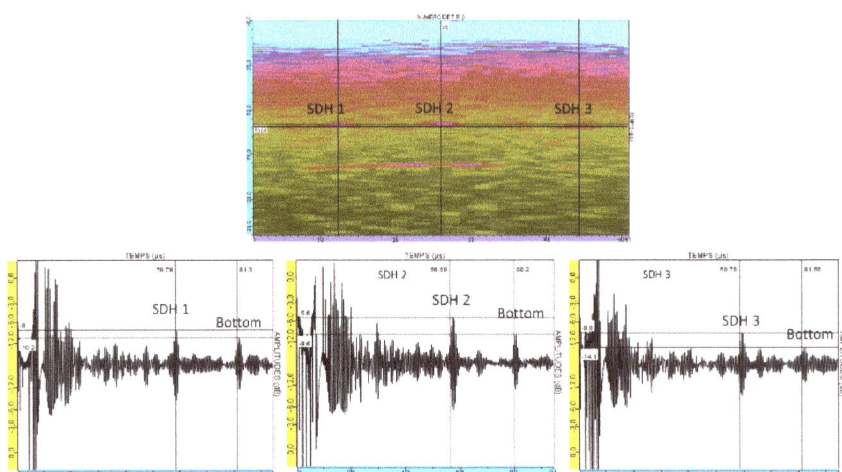

Figure 18. Experimental results. B-scans and A-scans of the 3 SDHs: A high pass filter was applied to eliminate low-frequency noise.

Between the SDHs and the backwall echoes, other echoes were observed that are not reproduced in the simulation of this experimental test (Figure 19, note that CIVA does not simulate the echoes observed in the dead zone). They were not related to transversal waves that might appear in the aluminum blocks as, if so, these echoes would be present in the simulated images when CIVA computes the T-wave contributions.

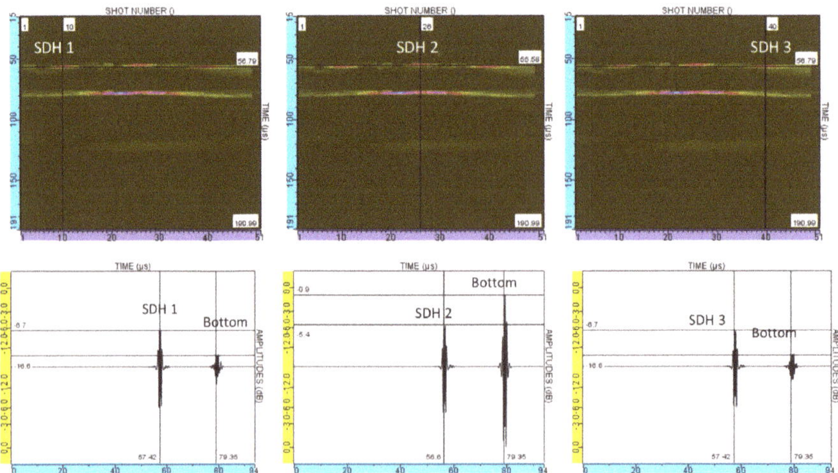

Figure 19. Simulation results with CIVA software. B-scans and A-scans of the 3 SDHs in the 250 mm aluminum block described in.

The experimental measurements and simulation results correspond: the SDH echoes and the backwall echoes are detected during the same range of time. However, it can be noticed that the dead zone is not taken into account in the CIVA simulation.

Under laboratory conditions, the results obtained on aluminum blocks with this new EMAT were improved compared to the results obtained with the previous EMAT (higher SNR and better deflection of the ultrasonic beam).

6. Conclusions

Phased array EMAT sensors offer a significant potential for use in liquid sodium, as illustrated with simulated and experimental results obtained with an EMAT prototype manufactured at the French CEA Non-Destructive Testing Department. The prototype's ability to steer and focus the ultrasonic beam to the desired focal spots using electronic delay laws was demonstrated. The low sensitivity of the EMAT described in this paper is still a drawback and a limitation to its application scope, despite the method proposed to improve it. However, laboratory tests showed a sensitivity high enough to image SDHs in aluminum blocks. Moreover, a comparison between the results obtained under laboratory conditions with the previous EMAT version (which provides some results in liquid sodium) and the new one shows the sensitivity improvement of the latter. Thus, good results in liquid sodium environments can be expected. The next step of this study is to perform tests under a liquid sodium environment at the Cadarache center of CEA by the end of 2019.

Author Contributions: Supervision, F.B.; Project administration, L.T.; Investigation, writing–review and editing, L.P. and R.R.

Acknowledgments: This work was supported through the 'Generation IV / Sodium Fast Reactors / ASTRID' tripartite program with CEA, EDF and FRAMATOME Company, funded by the French government.

Conflicts of Interest: The authors declare no conflict of interest. The funders had no role in the design of the study; in the collection, analyses, or interpretation of data; in the writing of the manuscript; or in the decision to publish the results.

References

1. Baqué, F.; Lhuillier, C.; Bourdais, F.; Navacchia, F.; Saillant, J.F.; Marlier, R.; Augem, J.M. R&D status on in-sodium ultrasonic transducers for ASTRID inspection. In Proceedings of the International Conference on Fast Reactors and Related Fuel Cycles: Next Generation Nuclear Systems for Sustainable Development (FR17), Ekaterinburg, Russia, 26 June 2017.
2. Saillant, J.F.; Marlier, R.; Navacchia, F.; Baqué, F. Ultrasonic transducer for non-destructive testing of structures immersed in liquid sodium at 200 °C. *Sensors* **2019**, *19*, 4156. [CrossRef] [PubMed]
3. Le Jeune, L.; Raillon, R.; Toullelan, G.; Baqué, F.; Taupin, L. 2D ultrasonic antenna system for imaging in liquid sodium. *Sensors* **2019**, *19*, 4344. [CrossRef] [PubMed]
4. Bourdais, F.; Marchand, B. Design of EMAT Phased Arrays for SFR Inspection. In Proceedings of the 40th Review of Progress in Quantitative NDE, Baltimore, MD, USA, 26 July 2013.
5. Bourdais, F.; Marchand, B. Experimental Validation of an 8 Element EMAT Phased Array Probe for Longitudinal Wave Generation. In Proceedings of the 41th Review of Progress in Quantitative NDE, Boise, ID, USA, 25 July 2014.
6. Bourdais, F.; Pollès, T.; Baqué, F. Liquid sodium testing of in-house phased array EMAT transducer for L-wave applications. In Proceedings of the ANIMMA 2015, Lisbon, Portugal, 20–24 April 2015.
7. Extende. Available online: http://www.extende.com/ (accessed on 9 October 2019).
8. Hilton, J.E.; McMurry, S.M. An adjustable linear Halbach array. *J. Magn. Magn. Mater.* **2012**, *324*, 2051–2056. [CrossRef]
9. Jian, X.; Dixon, S.; Edwards, R.S.; Morrison, J. Coupling mechanism of an EMAT. *Ultrasonics* **2006**, *44*, 653–656. [CrossRef] [PubMed]

Article

2D Ultrasonic Antenna System for Imaging in Liquid Sodium

Léonard Le Jeune [1,*], Raphaële Raillon [1], Gwénaël Toullelan [1], François Baqué [2] and Laura Taupin [1]

[1] French Alternative Energies and Nuclear Energy Commission—Laboratory for Integration of Systems and Technology CEA-LIST, 91191 Gif-sur-Yevtte Cedex, France; Raphaele.RAILLON@cea.fr (R.R.); gwenael.toullelan@cea.fr (G.T.); laura.taupin@cea.fr (L.T.)

[2] French Alternative Energies and Nuclear Energy Commission—Division of Nuclear Energy CEA-DEN, 13108 Saint-Paul-lez-Durance Cedex, France; francois.baque@cea.fr

* Correspondence: leonard.lejeune@cea.fr

Received: 30 August 2019; Accepted: 3 October 2019; Published: 8 October 2019

Abstract: Ultrasonic techniques are developed at CEA (French Alternative Energies and Nuclear Energy Commission) for in-service inspection of sodium-cooled reactors (SFRs). Among them, an ultrasound imaging system made up of two orthogonal antennas and originally based on an underwater imaging system is studied for long-distance vision in the liquid sodium of the reactor's primary circuit. After a description of the imaging principle of this system, some results of a simulation study performed with the software CIVA in order to optimize the antenna parameters are presented. Then, experimental measurements carried out in a water tank illustrate the system capabilities. Finally, the limitations of the imaging performances and the ongoing search of solutions to address them are discussed.

Keywords: SFR; in-service inspection; imaging; ultrasonic transducer; NDT (Non Destructive Testing); NDE (Non Destructive Evaluation)

---

1. Introduction

Since France considered in 2008 that the sodium-cooled reactor (SFR) concept was the most mature for Generation IV nuclear reactors, an extensive R&D programme was launched. In-service inspection was identified as a difficult task to be performed (as the sodium coolant is opaque, hot, highly chemically reactive, and difficult to drain). Ultrasonic techniques have been extensively studied as they are well adapted to Non Destructive Evaluation (NDE) and telemetry measurement in this harsh environment.

Thus, development of ultrasonic transducers to be immersed in sodium at about 200 °C in the reactor block (inspection is performed at shutdown conditions) led to a first phase of specification consolidation; then, a prequalification process involving increasingly more realistic experiments using acoustic techniques and simulations was performed with the patented CIVA code [1].

Associated applications for inspection deal with telemetry, vision, and volumetric control for SFR reactor block systems, structures, and components and for the power-conversion system.

Some techniques deal with short-distance imaging such as the EMAT (Electromagnetic-Acoustic Transducer) of CEA [2] and TUCSS (Ultrasonic transducer for Non Destructive Testing—NDT—of structures immersed in liquid sodium at 200 °C) of FRAMATOME [3] that are being developed for NDE, whereas other, such as the orthogonal imaging concept based on an underwater imaging system [4] and developed in the 2000s [5], deals with long-distance vision and the exploration of a large area without moving the probe. The orthogonal imaging system allows for scanning a much larger area than that scanned with the same number of elements arranged in a matrix pattern, and thanks to the large antenna's aperture in the focusing plane, remote targets are likely to be imaged. This concept has been taken up recently in order to reexplore its capabilities using actual simulation tools and imaging methods. The first simulated results and experimental measurements performed in water highlight the potential of the system in which manufacturing for sodium trials is ongoing at Toshiba Corporation while imaging algorithms are specifically adapted or developed for its applications.

2. Imaging Principle of the 2-Antenna System

2.1. Description of the 2-Antenna System

The 2-antenna system is made up of two identical linear-phased arrays disposed in a "T" arrangement (Figure 1a) which allows three-dimensional imaging. One antenna operates in emission (E; the red one) and the other in reception (R; the blue one). The surface of each element of the arrays is planar in the focusing plane and convex with a radius of curvature R in the orthogonal plane of the antenna (Figure 1a,b). The values of the antenna parameters are provided later.

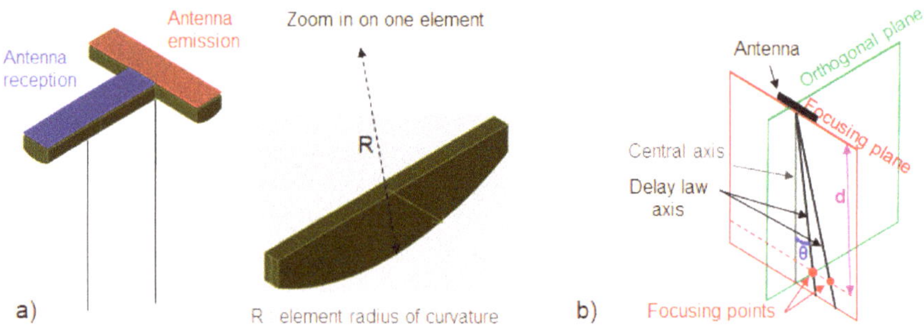

Figure 1. Diagram of the antennas' arrangement and definition of the main terms used in the article: (**a**) Diagram of the two antennas' arrangement and of the orthogonal section of an element and (**b**) definition of the key terms for one antenna. The central axis is perpendicular to the probe surface at its centre. The delay law axis represents the direction of the deflected beam when delay law is applied. When a delay law is applied, the beam is focused at the focusing point placed at a distance *d* as defined in the figure and or deflected by an angle θ in the focusing plane of the antenna. The orthogonal plane is perpendicular to the focusing plane.

2.2. Imaging Using Electronic Scanning

A delay law is applied to each antenna in order to focus its beam at a chosen focusing point in its focusing plane (Figure 2a,b) and, thus, to obtain a very narrow beam in this plane, whereas, due to the convex surface of the elements, the beam is very divergent in its orthogonal plane. Due to its shape, the beam of each antenna is called a "fan-beam" [5].

The intersection of the two fan-beams creates the useful part of the beam in the imaging process, named "cigar-beam" because of its shape, and is oriented along an axis located between the E delay law axis and the R delay law axis (Figure 2c). Thus, by applying successive delay laws in order to move the "cigar-beam", it is possible to scan and image a 3D zone (Figure 2d).

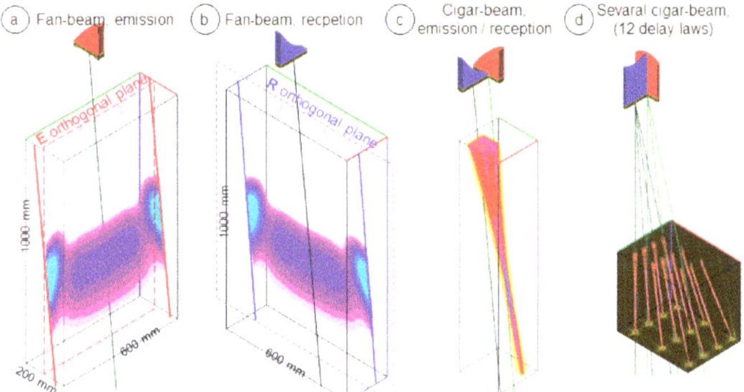

Figure 2. Simulated beam results: Illustration of the electronic scanning principle by visualization of the antenna beams computed with CIVA in a 3D zone of which the dimensions are given on the figure. (**a**) Fan-beam radiated by the emission antenna. (**b**) Fan-beam radiated by the reception antenna. (**c**) Emission/reception (E/R) cigar-beam. (**d**) Illustration of the twelve cigar-beams obtained for twelve delay laws (the beams are displayed together in the same image, but in reality, the twelve delay laws are applied successively to scan the space).

2.3. Imaging Using a Full Matrix Capture (FMC) and the Total Focusing Method (TFM)

To image the 3D area in front of the antennas, it is also possible to use the total focusing method after an FMC acquisition [6]. In this case, the FMC is performed by firing each element of the transmitting probe (E) and by receiving on every element of the receiving one (R). This imaging method is available in CIVA, and a simulated example of TFM applied for spherical targets is displayed Figure 3.

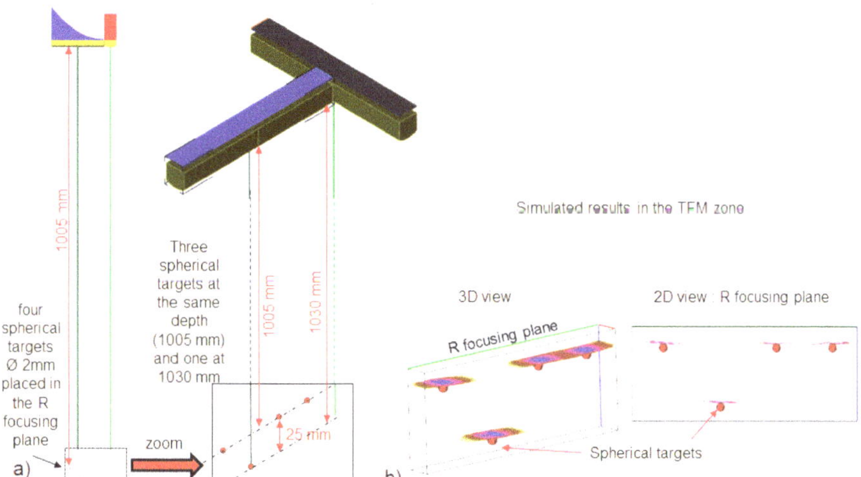

Figure 3. Simulated Full Matrix Capture (FMC)/Total Focusing Method (TFM) results: (**a**) Position of the spherical targets in front of the two antennas. The four spheres are in the R focusing plane (left); three of them are at a 1005-mm depth and one is at a 1030-mm depth (right). (**b**) Three-dimensional and 2D TFM images of the spherical four targets computed in a 3D zone placed around the targets as shown in Figure 3a.

2.4. Main Advantages and Drawbacks of the Imaging System

As said in the introduction, the main advantage of the 2-antenna system is the large aperture of each antenna, much larger than that obtained with the same number of elements arranged in a matrix pattern, allowing the improvement of the spatial resolution (see, for example, numerical values of the spatial resolution in Section 3.2). The main drawback comes from the way the cigar-beam is formed from the two fan-beams. Indeed, as the combination of the two fan-beams leads to a deflected cigar-beam of which the orientation depends on the position of the focusing point in the field of vision, the system is suitable only for targets generating specular echoes that come back to the receiver regardless of their orientation and position in the field of vision (as spherical targets or targets with rounded parts) and for targets, as the large, rough, curved surfaces of a reactor vessel, that diffract the field in all directions (that is the main application presented in Reference [5]). A distortion could be observed for targets generating specular echoes that do not come back to the receiver, but some targets could also be not detected at all if the specular echoes completely miss the receiver. For targets for which low diffraction echoes are involved in the imaging process, a small signal-to-noise ratio (SNR) is obtained. These considerations on the orientation of the cigar-beam used for the scanning imaging are also valid for the elementary beams impacting the target in the FMC/TFM imaging. Thus, the elementary echoes of the FMC might be invisible and lost in the noise, with the total target echo emerging from the noise only after the TFM computation.

3. Simulation Study for Antenna Parameter Optimization

Whatever the imaging method applied (scanning or TFM), the same requirements must be fulfilled regarding the radiated beam of the system to ensure an effective imaging in the case of the targeted applications (mainly far-field imaging of lost objects and detection of structure displacement). These requirements lead to the same optimized values of the antenna system parameters for both imaging methods.

3.1. Required Beam Features and Associated Influencing Parameters

The first requirements deal with the dimensions of the field of vision. In the orthogonal plane of each antenna, an almost constant beam's amplitude is required at each depth. In the direction of the focusing plane of each antenna, the potential grating lobes should be avoided. For depth, the amplitude should not decrease too quickly as long-distance imaging is contemplated. Other requirements concern the focal spot of the cigar-beam (axial and longitudinal resolutions).

The antenna parameters impacting these beam features are the classical phased array parameters such as the element size in the focusing plane, the gap between two adjacent elements, and the number of elements (these three parameters define the probe total aperture in the focusing plane) and also the antenna element's curvature and aperture in the orthogonal plane.

The effects of all these parameters in various combinations were evaluated and optimized with the CIVA software in relation to both the signal centre frequency fixed between 1 MHz and 2 MHz according to previous propagation studies in liquid sodium and to the bandwidth imposed by manufacturing constraints (about 30%). Some results of the CIVA study are presented in the next paragraph.

3.2. Example of the Effects of Some Influencing Parameters on the 2 Antenna Beams

The element's surface curvature and length in the orthogonal plane affect the beam divergence. For example, we present the study on the effect of the surface curvature. The simulated beams of one antenna (Figure 4) show that, when the element surface becomes more and more curved, from a flat surface to a curved surface with R = 30 mm, the beam divergence increases in the orthogonal plane, leading to a decrease in the radiated amplitude at a given depth. Thus, the chosen curvature was a trade-off between the divergence—required to ensure at each depth a constant amplitude in a large angular aperture—and the sensitivity.

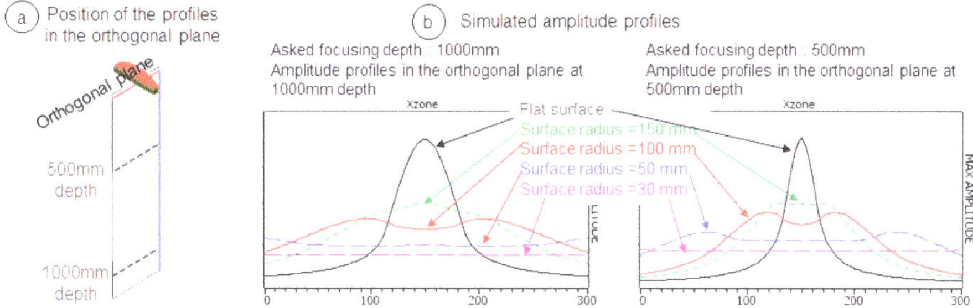

Figure 4. Simulated beam results: Effect of the variation of the radius of curvature of the elements in the orthogonal plane on the radiated beam at two depths. (**a**) Position of the profiles (dash lines) where the beam was computed. (**b**) Simulated amplitude of the beams along the 2 profiles obtained for different radii of curvature of the elements. The fixed parameters used for the computation were 128 elements for each antenna, with element's lengths of 1.8 mm in the focusing plane and of 35 mm in the orthogonal plane, a gap of 0.2 mm between adjacent elements, a centre frequency of 1.6 MHz, and a bandwidth of 30%.

The antenna aperture affects the spatial resolution as illustrated in Figure 5. As expected, the larger the aperture, the better the spatial resolution. To increase the antenna aperture, we can increase the element length, the pitch, or the number of elements. The element's length and the gap (i.e., the pitch) are imposed at the operating centre frequency as their increase leads to the presence of grating lobes. Then, we choose to increase the number of elements knowing that it will be limited by the maximum number of elements that can be operated and by manufacturing constraints. We hoped to build 128 elements antennas, but we know now that the number of elements will be comprised between 64 and 75.

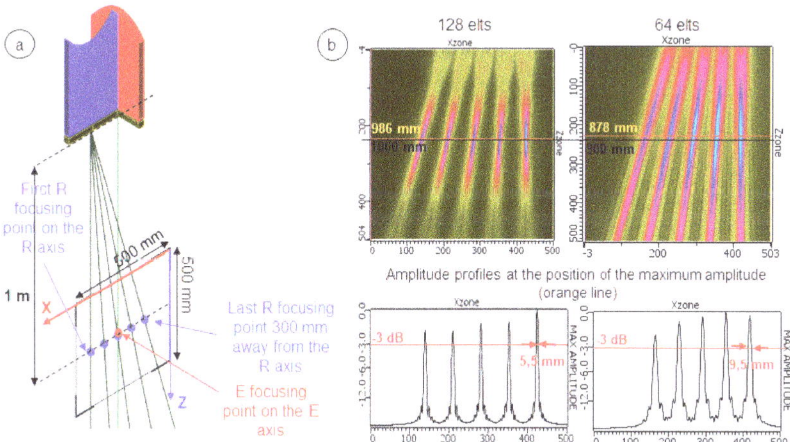

Figure 5. Simulated beam results: Effect of the variation of the antenna's aperture in the focusing plane on the cigar-beam spatial resolution at a given depth. (**a**) Position of the 2D beam computation area. (**b**) Simulated maximum amplitude of the beams obtained in this area for antennas with 64 elements (i.e., for an aperture of 127.8 mm) and with 128 elements (i.e., for an aperture of 255.8 mm). The fixed parameters used for the computation were element's lengths of 1.8 mm in the focusing plane and of 35 mm in the orthogonal plane, a gap of 0.2 mm between adjacent elements, a surface radius of curvature of 30 mm, a centre frequency of 1.6 MHz, and a bandwidth of 30%.

For example, with 128 elements (i.e., with an aperture of about 256 mm), when a delay law is applied to focus the emitted beam of one antenna at a point placed at a 1000-mm depth on its central axis, the focal spot dimension in its focusing plane is 5.5 mm. While, with 64 elements (i.e., with an aperture of about 128 mm) and the same point of focalization, the focal spot dimension increases to 9.5 mm (see Figure 5, bottom right).

Eventually, taking into account the simulation study performed with CIVA to optimize the antenna parameters for 3D imaging in liquid sodium at 200 °C leads to the following set of values for the antenna parameters. The number of elements is 128, the element length in the focusing plane is 1.8 mm, the gap between two adjacent elements is 0.2 mm, the element dimension in the orthogonal plane is 20 mm, the radius of the convex surface in the orthogonal plane is 30 mm, the centre frequency is 2 MHz, and the bandwidth of the excitation signal is about 30% with respect to the centre frequency.

4. Experimental FMC/TFM Images Performed in Water and Associated Simulations with CIVA

A two-antenna prototype was designed in 1999 by CEA and manufactured by IMASONIC French company to perform tests in a water tank using large curves and rough surfaces [5]. The number of elements of each antenna is 128, the element length in the focusing plane is 1.8 mm, the gap between two adjacent elements is 0.2 mm, the element dimension in the orthogonal plane is 20 mm, the radius of the convex surface in the orthogonal plane is 35 mm, the centre frequency is 1 MHz, and the bandwidth of the excitation signal is 50%. The relative positions of the two antennas are given in Figure 6.

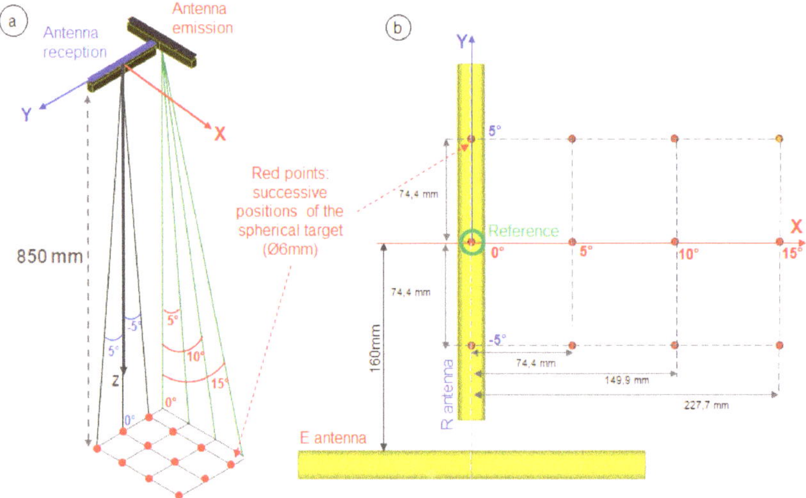

Figure 6. Diagram of the experimental trial configuration: (**a**) 3D diagram of the two-antenna system and successive positions of the spherical targets at a 850-mm depth (red points): the spherical target's FMC were measured one after the other for each of the twelve positions of the sphere. (**b**) Top view of the same configuration. The sphere located at X = 0° and Y = 0° (surrounded by a green circle) is used as a reference for the amplitudes (see in the text).

Various experimental FMC and TFM images of spherical and planar targets were performed in order to assess the antennas' capabilities in terms of sensitivity and spatial resolution. The associated FMC and TFM were also computed with CIVA (the model used for the echoes computation of the spherical targets presented below was the SOV (Separation of Variable) model). For example, some FMC acquisitions of the echoes of a spherical target (diameter Ø = 6 mm) located at a 850-mm depth were carried out for 12 different positions of the target in the antennas field of vision (Figure 6).

The TFM measured images obtained for four positions of the sphere are presented Figures 7 and 8. They show the good capability of system imaging as, even for the target far from the R antenna axis, the TFM image's quality is good in terms of SNR and resolution in the three cases of Figure 7a,b and Figure 8a. It is only when the sphere is far from both the E and R antenna axes (Figure 8b, where the sphere is distant 235 mm from the E antenna and 227 mm from the R antenna) that the SNR becomes too low to allow a good experimental image of the sphere. The SNR values measured on the experimental TFM are 9.5 dB for Figure 7a, 8 dB for Figure 7b, 8 dB for Figure 8a, and 2 dB for Figure 8b. It should be noted that, to calculate these SNRs, both attenuation and grating lobes were included in the noise measurement.

The four results also show a good qualitative agreement between CIVA and the measures regarding the evolution of the target image with the position of the sphere in the field of vision. A good quantitative agreement was also obtained for the maximum amplitude of the TFM images: the discrepancy between CIVA and the twelve measures was less than 3 dB (the reference for the amplitude is the TFM amplitude obtained for the sphere located at X = 0° and Y = 0°, which is surrounded by a green circle in Figure 6). However, the model used for the FMC simulation in CIVA does not predict the noise observed on the experimental image as it is not taken into account for the computation.

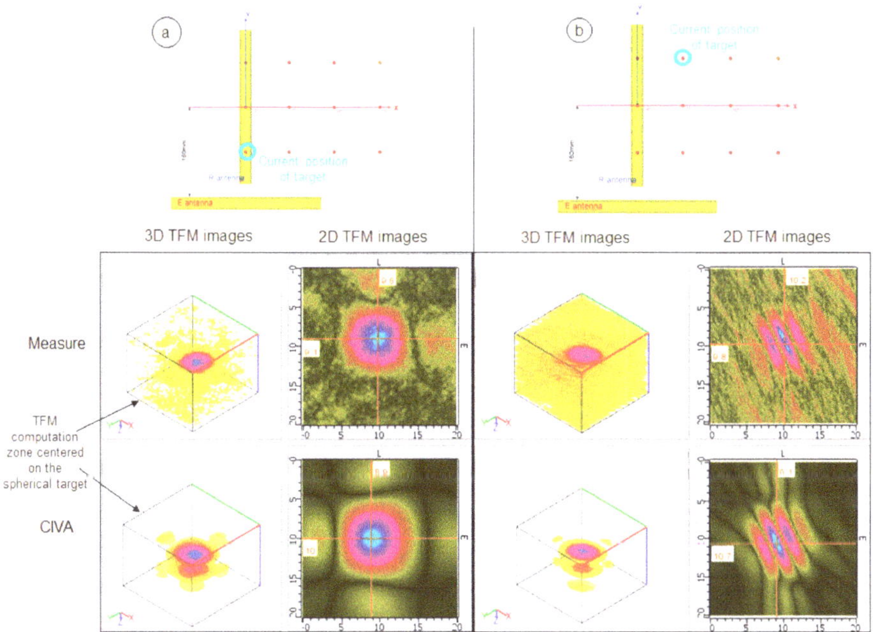

Figure 7. Measured and simulated FMC/TFM results: (**a**) The spherical target is at the position surrounded with a blue circle in the 2D diagram of the experimental trial configuration (top of the figure). (**b**) The spherical target (Ø: 6 mm) is at the position surrounded with a blue circle in the 2D diagram. For each position of the target, the measured and CIVA simulated 3D and 2D TFM images are displayed. The 3D TFM image was computed in a 3D zone centered on the spherical target and represented on the figure. The 2D TFM image is in the XY plane and was extracted from the 3D image at the position of the maximum amplitude.

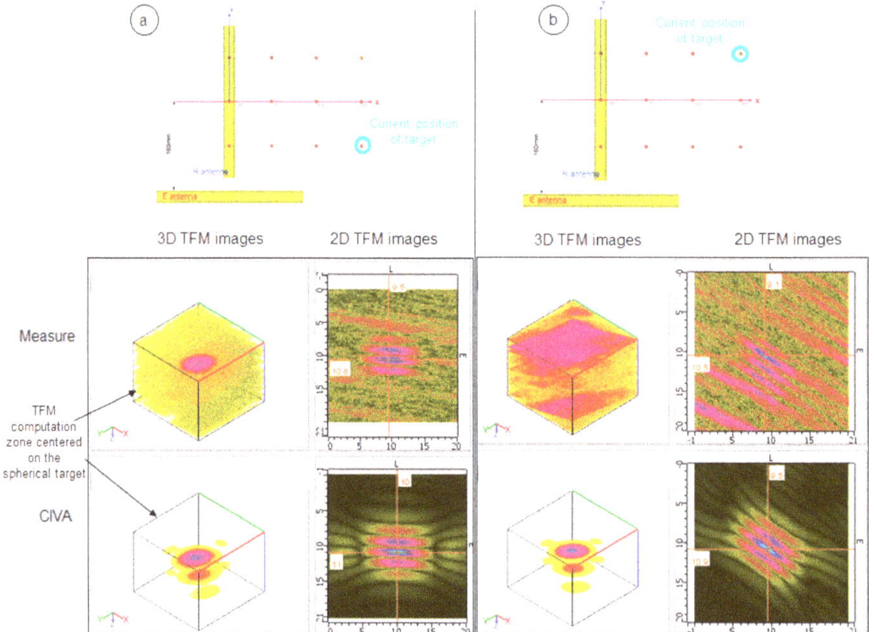

Figure 8. Measured and simulated FMC/TFM results for targets placed far from the antenna: (**a**) The spherical target is at the position surrounded with a blue circle in the 2D diagram of the experimental trial configuration (top of the figure). (**b**) The spherical target (Ø: 6 mm) is at the position surrounded with a blue circle in the 2D diagram. For each position of the target, the measured and CIVA simulated 3D and 2D TFM images are displayed. The 3D TFM image was computed in a 3D zone centered on the spherical target and represented on the figure. The 2D TFM image is in the XY plane and was extracted from the 3D image at the position of the maximum amplitude.

5. Discussion for Future Work: Improvement of the Images' Signal-to-Noise Ratio and their Computation Time

It has been experimentally shown that the TFM applied to an FMC acquisition performed with the two-antenna system allows to image spherical targets at large distances in water. However, two main drawbacks have been identified. The first one is the low Signal-to-Noise Ratio (SNR), especially for far-reaching targets. The second one is the high computation time required to obtain the images, which is due to the large size of the field of vision (several hundred centimetres) and the large number of elements of the antennas. In order to optimize the SNR and the computation time, several leads are being investigated.

5.1. Signal-to-Noise Ratio Improvement

During full matrix capture, each probe's element of the antenna operating in emission is fired individually and emits a spherical wave that propagates in the medium and gives rise to an elementary echo when meeting a target. This emission process, while it allows to insonify a large volume, transmits a limited acoustic power as only one element is used. In order to increase the acoustic power sent into the medium, the Plane Wave Imaging (PWI) method [7] can be used. The principle of this method is to emit plane wave fronts generated by all the elements of the probe at different angles instead of the spherical elementary wave front emitted by one element. As all the probe's elements are excited together, the acoustic power sent in the medium is much higher. Then, for the FMC, the backscattered signals are recorded on every element of the probe and the signals are coherently summed up to obtain an image similar to the one obtained by TFM. The sole difference when applying the TFM is the computation of the incident wave's time of flight (spherical for the FMC and plane for PWI). In order to take into account cylindrical wave fronts emitted by the element curved surface in the orthogonal plane, some modifications were added to the "classical" PWI method that deals with emitted planar wave fronts. Another possibility to improve the SNR is to use encoded transmissions like Hadamard spatial codes [8]. With this method, all the probe's elements are excited simultaneously, but for every transmission, N electric signals applied to the N elements are multiplied by a combination of a +1/−1 coefficient. A coded matrix is thus formed. The equivalent impulse response matrix (i.e., FMC) can be obtained with a decoding operation. The Hadamard coded excitation increases the SNR by an $\sqrt{N}$ factor compared to a conventional FMC.

The improvements resulting from the application of these two methods (PWI and Hadamard) are not illustrated for our configuration of interest because it requires new experimental trials, where the FMC will be replaced by a PWI or Hadamard acquisitions. These acquisitions are not completed at the moment. Thus, instead of illustrating for telemetry applications (long-distance case), we illustrate the methods for a highly attenuating medium (short-distance case). In each case, a high-level noise is induced by attenuation (by the propagation distance or the high attenuation coefficient) and the same methods can be applied to reduce it. Figure 9 shows the images obtained with three methods (TFM, Hadamard, and PWI) for a side-drilled hole (diameter of 2 mm) at a 25-mm depth in a strongly attenuating High-Density Polyethylene (HDPE) component with a 5-MHz, 64-element probe. The noticeable increase in SNR with the Hadamard coding and PWI demonstrates the benefit that can be expected from these methods.

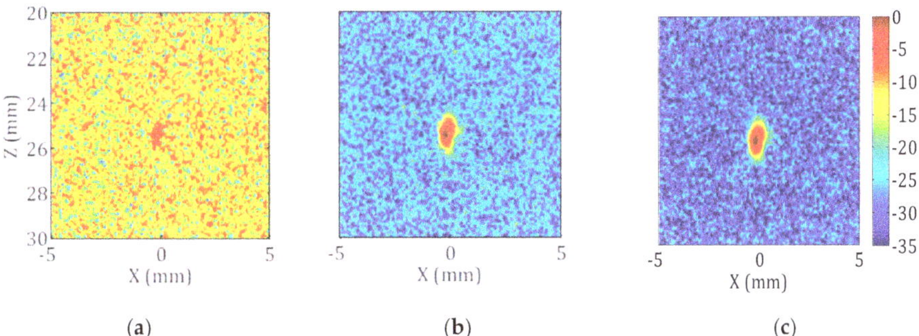

Figure 9. Comparison between (**a**) TFM, (**b**) Hadamard coding, and (**c**) Plane Wave Imaging (PWI) images obtained in a High-Density Polyethylene (HDPE) component for a 64-element, 5-MHz probe. The case is of a side-drilled hole (diameter of 2 mm) at a 25-mm depth. The amplification gain and voltage excitation are the same for the three acquisitions process. Attenuation of HDPE in the bandwidth is comprised between 0.4 and 1.4 dB/mm. Signal-to-Noise Ratios (SNR) are respectively 9 dB, 15 dB, and 18 dB [8].

5.2. Computation Time Improvement

The other axis for improvement is the computation time of the images. Three options are currently under investigation.

The first option is to use massive GPU (Graphic Processing Unit) parallelization. As the TFM, Hadamard, or PWI algorithms consist in a large number of operations (summations of amplitudes at a given time of flights for shot and receiving elements), they can be quite parallelized on a CPU (Central Processing Unit) or a GPU. In order to avoid poor scaling effects due to concurrent memory accesses, parallelization has been done on pixels and very good improvements in computation time have been obtained on both CPU and GPU architectures [9].

The second option, which can be used with any method and with parallelization, is to develop an "adaptive" grid in the imaged area. Here, the principle is to start the imaging process with a loose grid, then to detect the high-amplitude regions, and to iteratively refine the grid around those regions [10]. In our configuration that is purely three-dimensional, a 3D grid is needed. Work is ongoing to develop it, and at the moment, we can illustrate it only with a 2D configuration. Figure 10 shows 2D TFM images obtained on a set of simulated data for two side-drilled holes in a steel specimen imaged with 64 elements and a 2-MHz probe. It can be noted that, in this case, the same image quality is achieved by the adaptive grid with only 16% of the number of pixels used in the original case (area of 45×40 mm with a regular step of 0.4 mm).

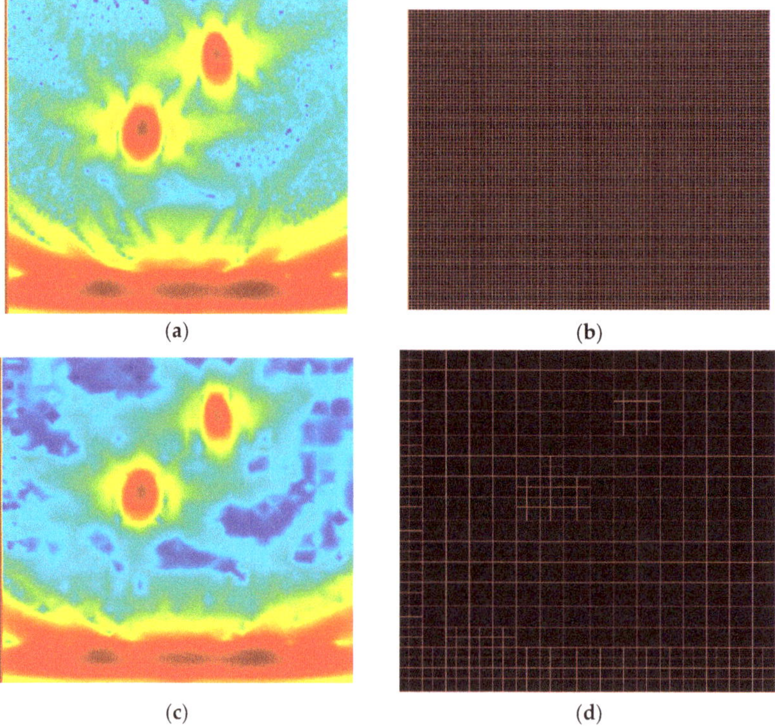

Figure 10. Adaptive grid: (**a**) TFM image obtained using (**b**) a regular and very fine grid and (**c**) TFM image obtained using (**d**) an adaptive grid. The number of pixels in Figure 10d represents 16% of the total number of points in Figure 10b.

The last option is to use frequency domain (f-k) methods. These methods are well known to greatly reduce computation times. In particular, the methods developed by Stolt and Lu has been adapted to NDE cases and to plane wave imaging [11]. It has been shown that f-k methods greatly reduce the algorithm complexity (number of operations) compared to the classical PWI (cf. Figure 11) and, thus, the computation time. The result presented Figure 11 was obtained in pulse-echo mode, and the work to extend it to our antenna system, with separate transmission and reception, is ongoing.

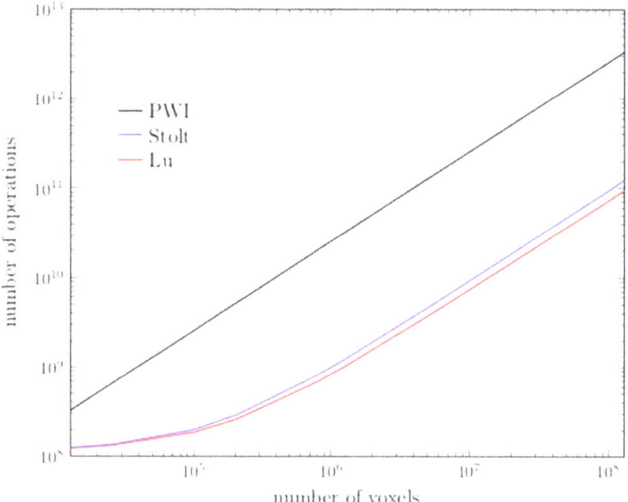

Figure 11. Algorithmic complexities: comparison between frequency domain (f-k) methods (Stolt and Lu) and PWI [11].

Furthermore, it is important to note that our configuration presents two advantages regarding the computation time of the images that strongly depends on the time required to compute time of flight between the elements and the points of the imaged area. The first one is the absence of interface between the elements of the antennas and the inspected area, and the second one is that the propagation occurs in a fluid medium (case of the sodium at inspection temperature), where only longitudinal waves exist. Indeed, the time-of-flight computation requires knowledge of the wave path between the transmitting element and the focusing point. This path is much easier and thus faster to compute when there is no interface as there is no refraction of the waves. Imaging the large area of the antenna field of vision is made possible thanks to these advantages combined with the implementation of the methods to decrease the image's computation time.

6. Imaging Acoustic Sensor for Sodium Conditions

This 2-antenna acoustic system is devoted to sodium viewing with the following applications: global imaging of the large primary circuit (some meters dimension), hypothetical large disorder detection (structure buckling), and potential lost-parts identification. The present paper exhibits the capabilities of such an enhanced imaging sensor which has been optimized with the CIVA software: experimental measurements were carried out in a water tank, and results show a good qualitative agreement between CIVA simulated results and the measures regarding spherical and planar target images. Having in hand this original concept of a 2-antenna acoustic system and the simulation CIVA code validated on commissioning water test results, it has been possible to design a first sodium prototype. Made of 64 curved elements for each antenna, it is being manufactured by Toshiba Corporation, which found and implemented specific technological solutions for bonding of the piezo material and backing material, front plate material, and array processing of the probe.

Commissioning tests of this prototype will be first performed in water in Japan by the end of 2019; then sodium tests are scheduled in France in 2020. Dealing with abovementioned applications for in-sodium vision, different targets will be used with different shapes: spheres for global viewing capacities of the antenna, flat surface with small basins to evaluate the antenna's performances for surface metrology application, and nuts for the lost-parts detection performances.

7. Conclusions

We presented a technique for long-distance vision in liquid sodium that enables imaging of large areas using a system of two orthogonal linear phased arrays (antennas) and the total focusing method. The use of CIVA to compute the antenna-radiated beams allows to easily understand and illustrate the effect of all the antenna parameters (frequency, aperture, pitch, etc.). CIVA also enables to study many combinations of parameters that was not possible in the first studies of this system and to propose an optimized set of values regarding the targeted application of long-distance viewing. We presented some examples of this simulated parametric study. The comparison of CIVA simulations with experimental results shows a good agreement aside from the noise that is not taken into consideration because CIVA does not compute it. The FMC/TFM imaging method was used to image the antennas' field of vision when, in the first study, the image was performed thanks to successive focalization at the different points of the field. This imaging method is very efficient as illustrated by the experimental results obtained in water, but as we deal with long-distance imaging and large imaging areas, small SNR and computation time appeared as limitations for this method. We proposed several leads to improve both of them that will be soon experimentally evaluated in water.

Author Contributions: Supervision, F.B.; Project administration, L.T.; Investigation, G.T. and R.R.; Writing–review and editing, R.R. and L.L.J.

Funding: This work was supported through the "Generation IV/Sodium Fast Reactors/ASTRID" program with CEA, EDF, and FRAMATOME funded by the French government. In the same framework, other ultrasonic transducers are being developed for NDE: TUCSS of FRAMATOME [3] and EMAT of CEA [2].

Acknowledgments: The authors would like to thank Toshiba Corporation for manufacturing the 2-antenna first prototype able to operate within liquid sodium at 200 °C in the frame of the CEA–Toshiba running agreement.

Conflicts of Interest: The authors declare no conflict of interest. The funders had no role in the design of the study; in the collection, analyses, or interpretation of data; in the writing of the manuscript; or in the decision to publish the results.

References

1. Extende. Available online: http://www.extende.com/ (accessed on 4 October 2019).
2. Pucci, L.; Raillon, R.; Taupin, L.; Baqué, F. Design of a phased array EMAT for inspection applications in liquid sodium. In Proceedings of the 58th Annual British Conference on Non-Destructive Testing, Telford, UK, 3–5 September 2019.
3. Saillant, J.J.; Marlier, R.; Navacchia, F.; Baqué, F. Ultrasonic transducer for non-destructive testing of structures immersed in liquid sodium at 200 °C. *Sensors* **2019**, *19*, 4156. [CrossRef] [PubMed]
4. Alais, P.; Cesbron, N.; Challande, P.; Ollivier, F. 3-D underwater imaging system. *Acoust. Imaging* **1995**, *21*, 723–728.
5. Imbert, C. Visualisation Ultrasonore Rapide sous Sodium, Application aux Réacteurs à Neutrons Rapides. Ph.D. Thesis, Claude Bernard Lyon 1 University, Lyon, France, 1997.
6. Holmes, C.; Drinkwater, B.W.; Wilcox, P.D. Post-processing of the full matrix of ultrasonic transmit–receive array data for non-destructive evaluation. *NDT E Int.* **2005**, *38*, 701–711. [CrossRef]
7. Le Jeune, L.; Robert, S.; Lopez Villaverde, E.; Prada, C. Plane wave imaging for ultrasonic non-destructive testing: Generalization to multimodal imaging. *Ultrasonics* **2016**, *64*, 128–138. [CrossRef] [PubMed]
8. Lopez Villaverde, E.; Robert, S.; Prada, C. Ultrasonic imaging in highly attenuating materials with hadamard codes and the decomposition of the time reversal operator IEEE Trans. *Ultrason. Ferroelectr. Freq. Control* **2017**, *64*, 1336–1344. [CrossRef] [PubMed]
9. Lambert, J.; Pédron, A.; Gens, G.; Bimbard, F.; Lacassagne, L.; Iakovleva, E. Performance evaluation of total focusing method on GPP and GPU. In Proceedings of the 2012 Conference on Design and Architectures for Signal and Image Processing, Karlsruhe, Germany, 23–25 October 2012.

10. Chouh, H. Simulations Interactives de Champ Ultrasonore pour des Configurations Complexes de Contrôle non Destructif. Ph.D. Thesis, Claude Bernard Lyon 1 University, Lyon, France, 2016.
11. Mérabet, L.; Robert, S.; Prada, C. 2-D and 3-D reconstruction algorithms in the Fourier domain for plane-wave imaging in non-destructive testing. *IEEE Trans Ultrason Ferroelectr Freq Control.* **2019**, *66*, 772–788. [CrossRef] [PubMed]

Article

Ultrasonic Transducer for Non-Destructive Testing of Structures Immersed in Liquid Sodium at 200 °C

Jean-François Saillant [1,*], Régis Marlier [2], Frédéric Navacchia [3] and François Baqué [3]

[1] INTERCONTROLE/Framatome, 4 Rue Thomas Dumorey, 71100 Chalon-sur-Saône, France
[2] Framatome, 10 rue Juliette Récamier, 69456 Lyon, France; regis.marlier@framatome.com
[3] C.E.A. (Commissariat à l'Energie Atomique et aux Energies Alternatives), Centre de Cadarache, DEN/DTN/STCP/LISM, 13108 St Paul Lez Durance, France; frederic.navacchia@cea.fr (F.N.); francois.baque@cea.fr (F.B.)
* Correspondence: jean-francois.saillant@framatome.com
† This paper is an extension version of the conference paper: Jean-François Saillant; Régis Marlier; François Baque. First results of Non-destructive testing under liquid sodium at 200 °C. In Proceedings of the 2016 IEEE International Ultrasonics Symposium (IUS), Tours, France, 18–21 September 2016.

Received: 28 August 2019; Accepted: 24 September 2019; Published: 25 September 2019

Abstract: TUCSS transducer (French acronym standing for Transducteur Ultrasonore pour CND Sous Sodium) is designed for performing NDT (Non-Destructive Testing) under liquid sodium. Under sodium, the tests results obtained show that these transducers have sufficiently good acoustic properties to perform basic NDT of a structure immersed under liquid sodium at about 200 °C using conventional immersion ultrasonic technics. Artificial defects were made next to an X-shaped weld and could clearly be detected.

Keywords: ultrasonic transducer; NDT; NDE; sodium; reactor; ultrasound; TUCSS; ISI&R; SFR; FBR

---

1. Introduction

Today, nuclear power provides about 10% of the world's electricity [1], principally generated from thermal–neutron reactors such as Pressurized Water Reactors (PWR) or Boiling Water Reactors (BWR). Parallel to the development of these thermal reactors, several countries have undertaken programs on Sodium Fast Reactor (SFR) development. To date, 12 experimental prototypes and six commercial-size reactors of SFRs have been designed, constructed, and operated [2]. SFRs offer great potential for a sustainable nuclear energy for many reasons, such as increased uranium utilization, fast nuclear power growth from the available resources through the breeding of fissile material, reduction of radioactive waste, utilization of fission products and other radioactive isotopes, higher thermodynamic efficiency.

Regardless of these advantages, SFRs pose specific problems related to the use of chemically reactive sodium. For instance, looking at the available feedback experience of SFRs operation, In-Service Inspection and Repair (ISI and R) is a significant issue to deal with and plays an important role in the safety approach of the plant for complying with new safety standards [3]. Non-destructive Testing (NDT) of the components immersed in liquid sodium is particularly challenging because of the harshness of the environment: high temperature, chemical aggression, and irradiation. The opacity of the medium is also a specificity, which makes ultrasonic testing a well-adapted candidate for performing NDT inside the primary circuit of these reactors.

This article reports on the feasibility of an ultrasonic transducer designed for performing NDT on a welded structure immersed in liquid sodium at approximately 200 °C. The design and fabrication of the ultrasonic transducer is first described, as well as the whole methodology for carrying out this specific ultrasonic inspection experiment. Finally, experimental results show that the developed transducer has sufficient acoustic properties to be considered for performing NDT during the outages of an SFR.

2. Ultrasonic Transducer for Non-Destructive Testing (NDT) under Liquid Sodium

Ultrasonic transducers for NDT under liquid sodium have to respond to very particular specifications. Technical difficulties to be overcome are first described before presenting the design of the Transducteur Ultrasonore pour CND Sous Sodium (TUCSS) transducer.

2.1. Technical Difficulties to be Overcome

2.1.1. Temperature Resistance

ISI and R of the reactor will be performed during outages, as sodium temperature is maintained at about 200 °C. This temperature is not conventional for ultrasonic transducers, and, moreover, the transducer has to withstand this temperature from several hours to several days, continuously. The transducer has to be functional at this temperature without external cooling, as it can be immersed under 10 m, or more, of sodium, which makes cooling of the transducer inefficient with such a depth.

Several technologies can be considered for generating ultrasound. The choice here was made to use piezoelectric transduction for converting electrical energy into ultrasonic pressure waves, and vice versa. If the piezoelectric crystal used is ferroelectric, it is important to pay attention to its Curie temperature so that it preserves its piezoelectric properties at the temperature of interest.

Passive materials conventionally used in NDT transducers fabrication cannot be used at 200 °C, as their physical integrity could be damaged. Moreover, thermal expansion coefficients of the materials of the different components (backing, matching layers, piezoelectric material, etc.) lead to different strains at interfaces, generating stresses which may cause bond failure and therefore malfunction of the transducer.

Electrical connections are also affected by the temperature level. So, cabling and soldering have to be chosen carefully.

2.1.2. Chemical Compatibility

It was mentioned in the introduction that sodium was chemically reactive (like other members of the alkaline elements family, it is a powerful reducing agent). It is reactive to air and water but it can also be aggressive to different materials, so much care has to be taken at the external conditioning of the probe. For instance, epoxy resins are not all compatible with sodium. Obviously, all materials used for the external packaging of the probe must be compatible with sodium.

2.1.3. Irradiation Resistance

Transducers will be exposed to high levels of gamma irradiation (there is no neutron irradiation, as inspections occur during reactor outage). Indeed, the potentially inspected structures can be quite close to the reactor core, and thus activated by the neutron flux during power generation cycles. Our objective is to design a transducer capable of withstanding a cumulated dose up to 216 kGy, corresponding approximately to a one-month inspection campaign underneath the SFR core.

2.1.4. Acoustic Performances Compatible with Non-Destructive Testing (NDT) Requirements

Short ultrasonic pulses are necessary in order to have sufficient resolution for efficient NDT inspection. Controlling the pulse shape is necessary and our objective is to have a ringdown of a maximum of 4 cycles at −20 dB at operating resonant frequency. Also, the dynamics of the transducer have to be fairly high in order to be able to sense echoes from potential flaws inside the matter. In a first approach, we aim to have signal to noise ratio (SNR) > 50 dB on the entrance echo of a flat stainless-steel block.

The acoustic properties of sodium are relatively similar to that of water, even considering that sodium is a metal in its pure state. At 200 °C, its properties are [4]:

- Density = 903.5 kg/m^3
- Sound velocity = 2471.8 m/s
- Acoustic impedance = 2.23 MRayl

As a consequence, the same design rules can be applied as those for underwater immersion NDT transducer design. The problem mainly consists in finding suitable backing and matching layer materials for operating at a continuous temperature of 200 °C.

The most critical acoustic issue to be addressed is the acoustic transmission at the interface between sodium and the transducer's emissive face. Indeed, liquid sodium does not wet all material surfaces. By comparison, it has a similar behavior to a drop of mercury which remains more or less spherical when put in contact with a regular table, rather than spreading flat like a drop of water would do. When a liquid does not wet a surface, it means that there remains a thin film of gas in between. Transmission of acoustic energy between solid/gas or between liquid/gas is very inefficient in the MHz range; the acoustic transmission coefficient is in the order of 0.001%, compared to optimal state. This means the transducer's emissive face should be made from a material that can be wetted by sodium at 200 °C. Again, temperature can be a parameter of influence in the wetting of materials. For example, sodium does not wet stainless steel at 200 °C; however, it wets well at a temperature above 400 °C, as surface oxides react with sodium [5]. Note that when stainless steel wetting by sodium is achieved, it remains wetted even if temperature decreases.

2.2. Transducer Design and Fabrication

Several transducer technologies have been developed for under sodium viewing. Most of them are listed in the review paper [6–12]. EMAT (Electromagnetic acoustic transducer) technology also seems to be interesting, due to the fact that wetting is not necessary to generate or receive ultrasonic waves [13]. Nevertheless, NDT is very demanding in terms of acoustic performances because of the need to satisfy to both constraints of short pulse lengths and high signal dynamics. To date, and to the authors' knowledge, no transducer technology has proven capable of performing conventional NDT in immersion under liquid sodium at 200 °C.

TUCSS transducer (French acronym standing for Transducteur Ultrasonore pour CND Sous Sodium) has been designed and fabricated at INTERCONTROLE/Framatome. Its main objective is to demonstrate the feasibility of performing conventional NDT under the technical constraints listed in Section 2.1. A photograph of a TUCSS transducer is shown in Figure 1.

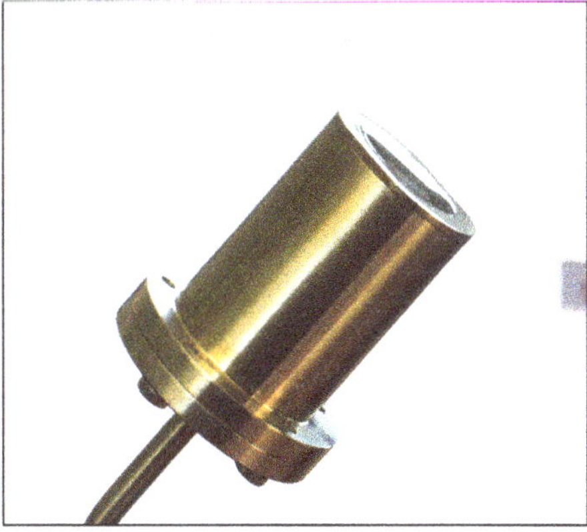

Figure 1. Photograph of a "Transducteur Ultrasonore pour CND Sous Sodium" (TUCSS) transducer.

TUCSS is based on piezoelectric transduction technology for generating and sensing ultrasonic waves. A schematic representation of the transducer is shown in Figure 2. Its characteristics are as follows:

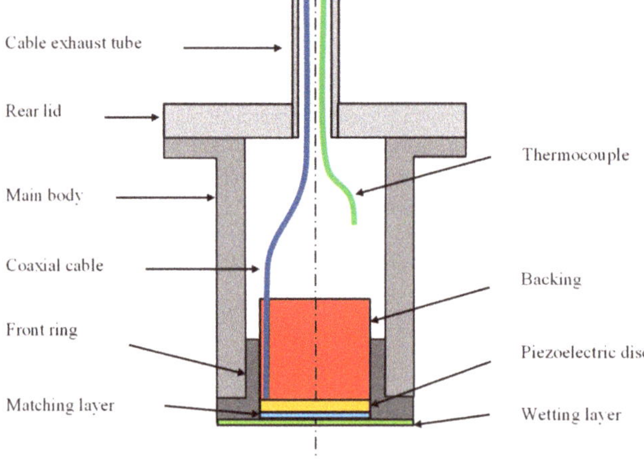

Figure 2. Schematic representation of the structure of a Transducteur Ultrasonore pour CND Sous Sodium (TUCSS) transducer.

The piezoelectric disc is made from a PZT (Lead Zirconate Titanate) ceramic NAVY type II that has a Curie point of 350 °C. Its free resonant frequency is 2.2 MHz and has an active diameter of 20 mm.

The backing material is made from a mixture of attenuative high temperature elastomers and large size particles. Its roles are to damp the vibration of the piezoelectric disc and to attenuate the acoustic energy travelling backwards (so that there are no spurious echoes originating from inside the transducer). Its thickness is approximately 20 mm.

The matching layer is made from low attenuation high temperature thermosetting resin. Its role is to optimize the transfer of acoustic energy travelling forward. Its thickness is 250 μm and its acoustic impedance at 200 °C is 2.53 MRayl.

The front layer used for wetting in sodium is made of elastomer. Its thickness is 70 μm and its acoustic impedance at 200 °C is 0.59 MRayl. Its properties were measured using the technique in Reference [14].

The cable is custom PFA (Perfluoroalkoxy) coaxial cable.

A type K thermocouple is integrated to monitor the temperature inside the transducer during testing and during pre-heating of the transducer before immersion (to avoid thermal shocking).

All the mechanical parts (main body, front ring, and rear lid) are made from 316 L stainless steel. The length from the inside face of the shoulder to front face of the transducer is 50 mm. The diameter of the body is 36 mm.

3. Materials and Methods

Prior work using TUCSS transducers [15] reported on experiments related to the "contact" NDT technique. The transducer was brought into contact with the testing block, as the whole setup was immersed in liquid sodium. The initial result was positive, as a ⌀ 3 mm side drilled hole (SDH) could be detected in the matter. However, it was found that the friction between the two contact surfaces was causing the loss of wetting on both surfaces (transducer and inspected block).

Further work using TUCSS transducers reported on "immersion NDT technique" [16], where the transducer was shifted away from the surface of the inspected part (i.e., not in contact). The results were satisfactory and it was decided to pursue this direction for the NDT of welded assemblies of SFR structures. Immersion NDT techniques are very common in the industry [17], and especially in the nuclear industry [18].

The present NDT experiment was conducted on a 316L stainless steel block. Here we considered a structure representing two 40 mm thick plates welded together by X-shape Manual Shielded Metal Arc Welding (Manual SMAW).

Austenitic welds add a great complexity to the NDT inspection of a structure. The different passes (see Figure 3) and the resulting recrystallized microstructures cause local heterogeneities in the stiffness matrix, which disturb and scatter the ultrasonic wave. The aim of the present study is to determine if it is possible to detect a flaw placed beyond the weld, in immersion under liquid sodium at 200 °C.

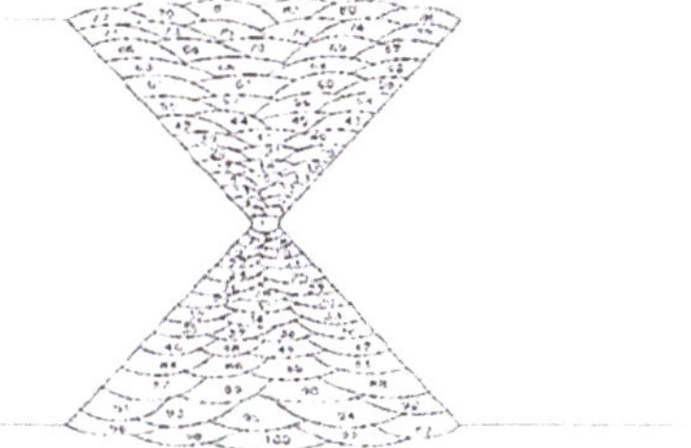

Figure 3. Multiple passes of the X-shape Shielded Metal Arc Welding (SMAW) weld.

The test block considered in this experiment is represented in Figure 4. It consisted of a 190 × 40 mm block of 316 L stainless steel, with a height of 100 mm. It can be seen that it included the X-shape weld and a 20 × 0.3 mm notch (with a notch height of 40 mm). This notch is made by spark machining from the back surface and is placed at mid-height of the block. Two other indications were added to the block in order to have reference points: a ⌀ 4 mm side hole (drilled all the way through the height of the block) and a 45° chamfer.

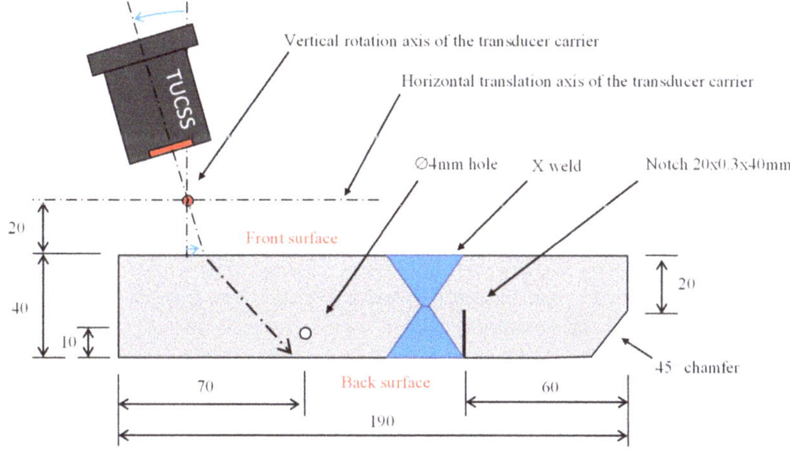

Figure 4. Schematic representation of the test block (horizontal cross section).

Under-sodium tests were conducted in a thermally regulated cylindrical vessel (⌀ 320 × 200 mm) containing 10 L of liquid sodium (pure sodium fusion temperature is 98 °C). A characterization device, called DEFO (Dispositif d'Essais Faisceau Oblique), was specifically designed and fabricated in order to accurately move the TUCSS in front of the test block inside this sodium filled vessel. Photographs of DEFO placed in the sodium vessel can be seen in Figure 5.

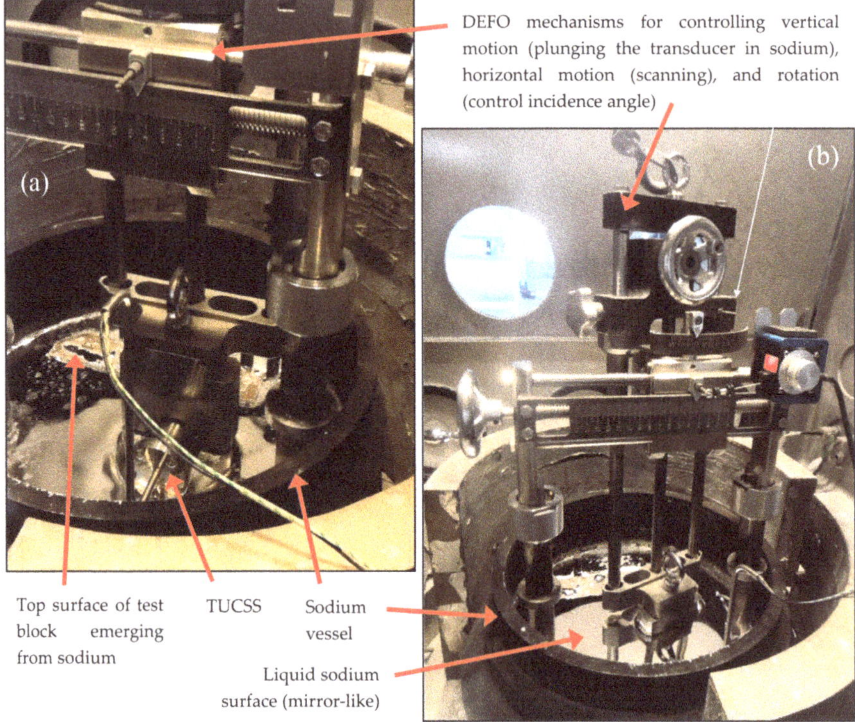

Figure 5. Photographs of experiment (**a**) before immersion of the transducer and (**b**) when the transducer is immersed.

Its mechanisms are made for translating and rotating the transducer and are kept as simple as possible. They allow the transducer to move according to the 3 following motions:

Vertical translation for plunging of the transducer into liquid sodium.

Horizontal motion along the translation axis (see Figure 4) for performing the ultrasonic scan.

Rotation around the vertical axis (see Figure 4) to control the incidence angle with respect to the front surface of the test block. This rotation applies to the transducer carrier and its axis is placed 20 mm away from the front surface. The carrier can keep its orientation when it is moved along the horizontal translation axis. The transducer's front face is placed 25 mm away from the rotation axis.

All the motions are manually controlled. Motors were excluded because of increased complexity brought by inert argon atmosphere (electrical arcing), presence of sodium aerosols, high temperatures, and instrumentation of the existing glove box. This experiment was conducted in a glove box of CEA-DEN (Cadarache, France) sodium facilities.

Acquisitions are made with a TabletUT ultrasonic electronic system (Mistras Eurosonic, Vitrolles, France). The electronic system was connected to a POSIWIRE WS42-1000-6-IE24LI-2-LF (ASM, Moosinning, Germany) wire coder for correlating ultrasonic data with positions along the horizontal scan.

Preparation of the experiment began by cleaning all the parts of the test device with alcohol, followed by complete drying at 120 °C for 24 h. The device was then entered into the glove box and placed into the sodium vessel. The volume of sodium was adjusted to reach the top surface of the test block. At this stage, the TUCSS transducer was not mounted on the DEFO device. The temperature of sodium then rose to 400 °C for 6 h. This temperature step is compulsory for the sodium to wet the stainless-steel test block and, consequently, for ultrasonic waves to be able to penetrate it (and to be representative of reactor conditions where all internal structures have been wetted by sodium during the initial high temperature cycle). Once the sodium cooled down to 200 °C, the TUCSS was mounted on the transducer carrier, still above the sodium surface, which was then translated down until it touched the surface of sodium (Figure 5a). This allows for a slow pre-heating of the transducer, thus avoiding thermal shocking. The transducer was plunged under the free surface of liquid sodium (Figure 5b) once its thermocouple read a stable temperature (approximately 170 °C). Echoes from the block were visible straight away, indicating that wetting was immediate. The system was then ready for performing ultrasonic scans.

4. Results

Several scans have been performed for different incidence angles. Each incidence angle allows the indications inside the test block to be seen with a different detection mode. We first present the normal incidence condition, which makes it easier to familiarize oneself with the reading of the data, and then we present the two incidence angle conditions for which best detection results are obtained.

4.1. Normal Incidence Condition

Figure 6 shows a scan done in normal incidence (i.e., TUCSS axis perpendicular to the surface of the test block).

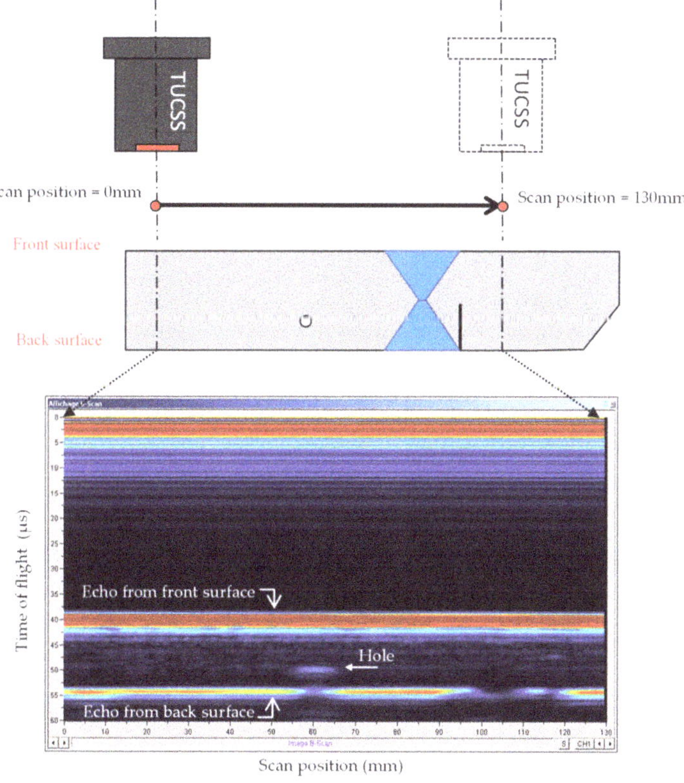

Figure 6. Normal incidence scan.

This scan can be interpreted as follows:

The trace from 0 to 12 µs is the saturated dead zone of the transducer.

The linear trace at 40 µs is the echo coming from the front surface of the block. It is present at all scan positions from 0 mm to 130 mm.

The spot trace at [60 mm, 50 µs] is the echo coming back from the ⌀ 4 mm hole.

The linear traces at 55 µs are echoes coming from the back surface of the block. This backwall echo is not visible at all scan positions:

- It is shadowed by the hole at position 60 mm,
- Between 95 mm and 110 mm, the ultrasonic beam is disrupted/attenuated by the weld, preventing echoes from coming back,
- Between 115–120 mm, the local disruption in the backwall echo is caused by the presence of the notch.

The settings for this scan were as follows:

- Sodium temperature = 200 °C.
- 2 MHz–100 V single square wave excitation.
- 30 dB gain in reception, in order to have the backwall echo without saturation (it was 12 dB for having the front face echo not saturated).

This scan shows that the weld has an important impact on the propagation of ultrasound, as the detection of the backwall echo is really disturbed. Also, even though a 0° incidence should not be used for detecting a notch made perpendicular to the back surface, this scan shows that the shadowing of the backwall echo gives us an indication of its presence.

4.2. Oblique Incidence Condition—Use of Shear 45° Waves

The first critical angle is reached at an incidence angle of 25.53°. Beyond this angle, the entire incident wave in sodium is converted into shear wave in the test block. The notch can be detected using pure shear wave mode. It was found that the best amplitude is obtained at an incidence angle of 35°, producing 45° shear waves (SW45°) in the test block. Figure 7 shows the resulting scan at 35° incidence angle.

The settings for this scan were as follows:

- Sodium temperature = 200 °C.
- 2 MHz–100V single square wave excitation.
- 45 dB gain in reception.

This scan can be interpreted as follows:

The traces from 0 to 20 µs are the saturated dead zone of the transducer. It is longer than the dead zone seen for the scan at normal incidence because the reception gain is 15 dB higher.

The traces at 45 µs all across the scan are echoes coming from the front surface of the block.

The trace at [5 mm, 70 µs] is the echo coming back from the ∅ 4 mm hole.

The 45° chamfer is represented by the linear trace going from [95 mm, 80 µs] to [115 mm, 70 µs].

Traces at [45 mm, 75 µs] to [48 mm, 85 µs] are echoes originating from the notch.

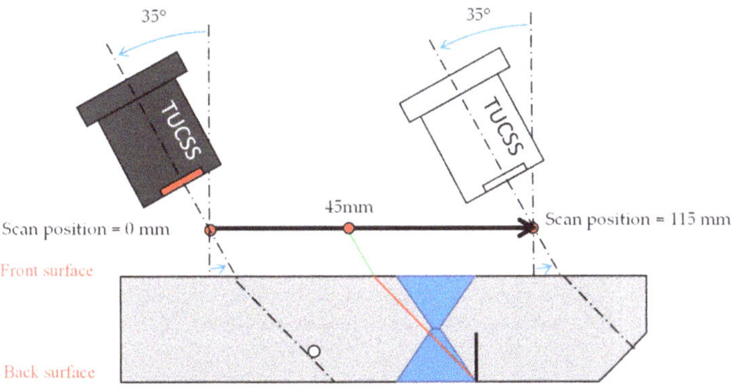

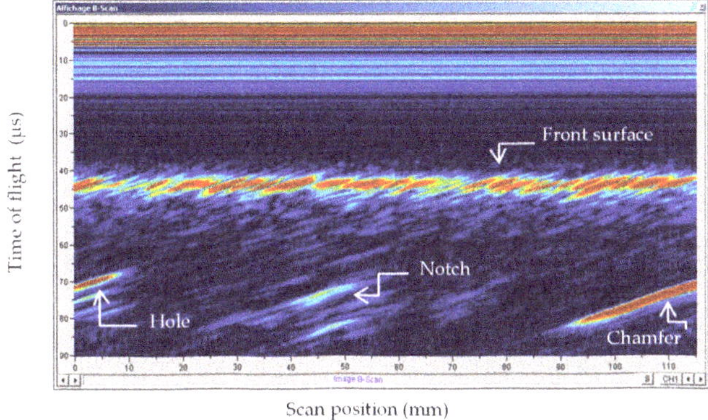

Scan position (mm)

Figure 7. Scan at an incidence of 35°, generating 45° shear waves in the block.

4.3. Oblique Incidence Condition—Use of LLS Detection Mode

When the incidence angle is lower than the first critical angle (25.53° in the present case), the incident wave is converted into two waves with different polarizations: a longitudinal wave (LW) and a shear wave (SW). These two waves have different propagation velocities and their refraction angles are therefore different (see Section 5 for wave velocities measurements). In this study, it was found that an incidence angle of 23° gave the best result for detecting the notch. At this angle, the two waves generated are LW65° and SW29°. Figure 8 shows the resulting scan at 23° incidence angle.

The settings for this scan were as follows:

- Sodium temperature = 200 °C.
- 2 MHz–100 V single square wave excitation.
- 45 dB gain in reception.

This scan can be interpreted as follows:

The traces from 0 to 20 µs are the saturated dead zone of the transducer. It is again longer than the dead zone seen for the scan at normal incidence because the reception gain is again 15 dB higher.

The traces at 40 µs all across the scan are echoes coming from the front surface of the block.

The trace at [20 mm, 65 µs] is the echo coming back from the ∅ 4 mm hole, detected in direct SW29°.

The notch is represented by the linear trace going from [25 mm, 80 µs] to [45 mm, 75 µs]. It is detected in LLS reflections mode. The first refracted LW65° is reflected off the notch in LW65°, which is then reflected off the backwall in SW29°. The resulting wave comes out in in sodium at the same position and angle (LW23°) as the incident wave.

The 45° chamfer cannot clearly be detected at this incidence angle. It produces some noise visible in 90 µs–115 µs range.

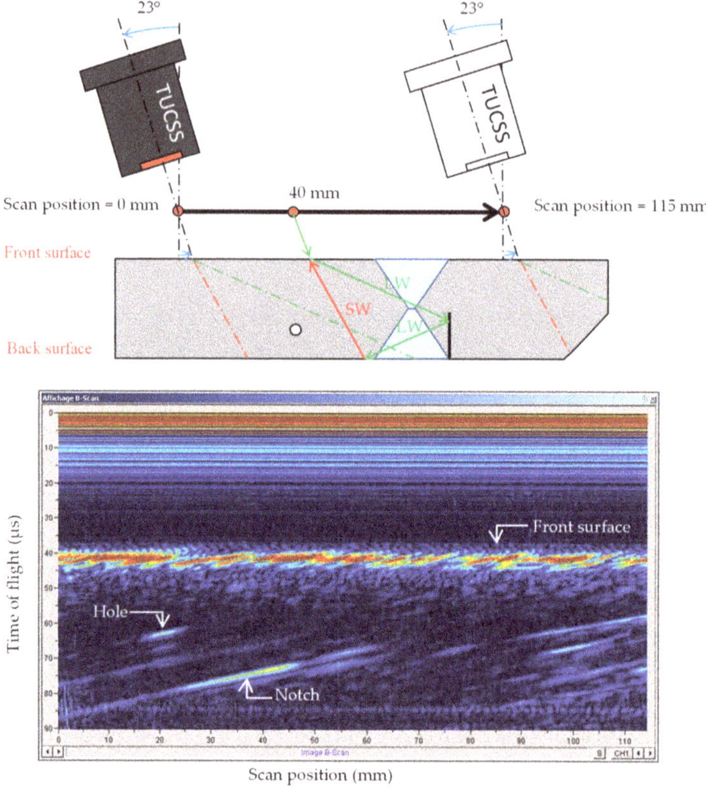

Figure 8. Scan at an incidence of 23°, generating LW65° and SW29° in the block.

5. Discussion

5.1. Acoustic Performances

The results shown in Section 4 clearly indicate that it is possible to detect a notch placed beyond the weld using a 2 MHz TUCSS transducer. Immersion testing was performed at different incidence angles, which allowed the notch to be detected with different modes. The best results were obtained using direct SW45° mode and LLS mode using an initial LW65°. In both modes, echoes from the notch were detected with a 45-dB gain for a 100 V excitation at 2 MHz. This is 33 dB higher than the gain required to detect the front surface echo without saturation.

The resolution of the transducer was also found to be sufficient. Echoes typically had an approximately 2-µs length, corresponding to a pulse ringdown of approximately 4 wavelengths at 2 MHz. Figure 9 shows a typical A-scan made with 0° incidence angle. The main echo at 40 µs is the echo from the front surface of the test block and the weaker echoes are echoes from the backwall.

These acoustic properties distinguish themselves from those of the other transducer by the resulting combination of high sensitivity and high damping. We attribute this to the fact that most of the transducers fabricated up to date for under sodium operation include a metallic front face. Stainless steel has an acoustic impedance of approx. 47 MRayl, while sodium has an acoustic impedance of 2.23 MRayl at 200 °C. Metallic front face is an acoustic barrier to the transfer of acoustic energy in liquid sodium. Here, we show that a solution based on elastomeric materials is physically feasible and acoustically relevant.

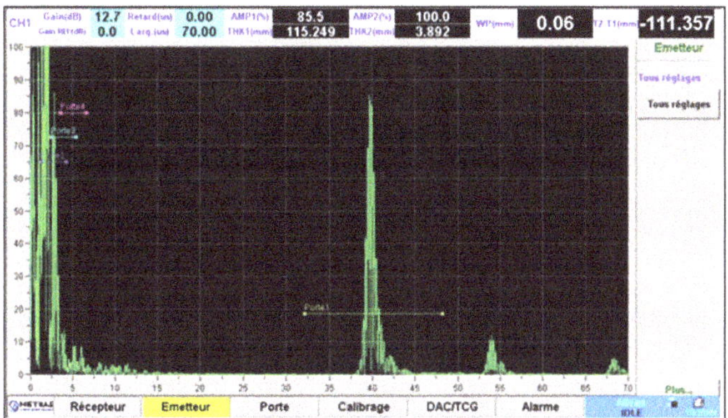

Figure 9. A-scan at incidence = 0°, scanning position = 30 mm, sodium temperature = 200 °C.

5.2. Measurement of Velocities

In our test conditions, sound velocity in sodium at 200 °C was measured at 2407 m/s (2.6% lower than theoretical value of 2471 m/s).

Sound velocities in 316 L stainless steel at 200°C were determined at:

- $V_{L\text{ in }200°C\text{ SS}}$ = 5584 m/s
- $V_{S\text{ in }200°C\text{ SS}}$ = 3004 m/s

5.3. Further Work

5.3.1. Time-Dependent Performances

The results shown in this article were reached during the first day of immersion of the transducer. It was noticed that performance slowly degraded day by day due to a loss of wetting of the front face. Nevertheless, the transducer can be repaired after reconditioning the elastomeric front face. Work is ongoing to understand the origin of this loss of wetting condition and to limit its effects.

5.3.2. Under-Sodium Tests with Mockups Including Narrow Gap TIG Weld

Further work should be pursued in a similar study using a mockup, including a weld made by automated narrow gap TIG (Tungsten Inert Gas) welding. Indeed, such a weld will present a different geometrical and microstructural environment for the propagation of ultrasonic waves and thus disrupt the detection of artificial defects in a different way than the present X-shape SMAW weld.

5.3.3. Irradiation

All the components and several material assemblies of the TUCSS transducers have been tested under irradiation up to 135 kGy without degradation. We can confidently assume that the whole transducer should be able to withstand 216 kGy (see specifications in Section 2), although a proper test under 200°C conditions is necessary to validate this point.

6. Conclusions

The experimental results obtained show that the new concept of TUCSS transducers exhibits sufficiently good acoustic properties to perform basic NDT of a structure immersed under liquid sodium at 200 °C, using conventional immersion ultrasonic technics.

The results obtained during this study demonstrate that conventional immersion ultrasonic NDT technics can be considered in this chemically aggressive environment during SFR outages. This work is an important step towards improved In-Service Inspection (ISI) of sodium-cooled nuclear reactors or other sodium related facilities.

Author Contributions: Transducer design, J.-F.S.; Fabrication and testing, J.-F.S.; Project coordination, R.M.; Organization and coordination, F.B. and F.N.

Funding: The work presented in this paper has been performed in the framework of the French Generation IV Program led by CEA with the support of the SFR tripartite R&D program, involving CEA, EDF and Framatome, and the ASTRID program through the collaborative agreement between CEA and Framatome. In the same framework, other ultrasonic transducers are being developed by CEA for in-sodium inspection: EMAT [19] and 2D antennas system for Under Sodium Vision [20].

Acknowledgments: The authors would like to thank Ciaran Verdelli (CEA) for all the work involving equipment manipulations inside the glove-box.

Conflicts of Interest: The authors declare no conflict of interest. The funders had no role in the design of the study; in the collection, analyses, or interpretation of data; in the writing of the manuscript, or in the decision to publish the results.

References

1. Energy, Electricity and Nuclear Power Estimates for the Period up to 2050. Available online: https://www-pub.iaea.org/MTCD/Publications/PDF/RDS-1-38_web.pdf (accessed on 24 September 2019).

2. Status of Fast Reactor Research and Technology Development. Available online: https://www-pub.iaea.org/MTCD/Publications/PDF/te_1691_web.pdf (accessed on 24 September 2019).

3. Jadot, F.; Baqué, F.; Jeannot, J.P.; de Dinechin, G.; Augem, J.M.; Sibilo, J. ASTRID Sodium cooled Prototype: Program for improving. In Proceedings of the 2011 2nd International Conference on Advancements in Nuclear Instrumentation, Measurement Methods and their Applications, Ghent, Belgium, 6–9 June 2011; pp. 3–37.

4. Sobolev, V. *Database of Thermophysical Properties of Liquid Metal Coolants for GEN-IV*; SCK-CEN: Mol, Belgium, 2011.

5. Paumel, K.; Moysan, J.; Chatain, D.; Corneloup, G.; Baqué, F. Modeling of ultrasound transmission through a solid-liquid interface comprising a network of gas pockets. *J. Appl. Phys.* **2011**, *110*, 044910. [CrossRef]

6. Griffin, J.W.; Bond, L.J.; Peters, T.J.; Denslow, K.M.; Posakony, G.J.; Sheen, S.H.; Chien, H.T.; Raptis, A.C. *Under-Sodium Viewing: A Review of Ultrasonic Imaging Technology for Liquid Metal Fast Reactors*; No. PNNL-18292; Pacific Northwest National Lab. (PNNL): Richland, WA, USA, 2009.

7. Tarpara, E.G.; Patankar, V.H.; Vijayan Varier, N. *Under Sodium Ultrasonic Viewing for Fast Breeder Reactors: A Review, Government of India Atomic Energy Commission*; Bhabha Atomic Research Centre: Mumbai, India, 2016.

8. Hans, R.; Kranz, E.; Weiss, H. Under sodium viewing—a method to identify objects in an opaque medium, Liquid metal engineering and technology. In Proceedings of the 3rd international conference on liquid metal engineering and technology, Oxford, UK, 9–13 April 1984.

9. Wang, K.; Chien, H.T.; Lawrence, W.P.; Engel, D.; Sheen, S.H. Ultrasonic Waveguide Transducer for Under-Sodium Viewing in SFR. *Int. J. Nucl. Energy Sci. Eng.* **2012**, *2*, 39–44.

10. Saruta, K.; Shirahama, T.; Yamaguchi, T.; Ueda, M. A consideration on the use of shear waves to improve the sensitivity of an optical ultrasonic sensor for under-sodium viewers. *e-J. Adv. Maint.* **2018**, *10*, 1–8.

11. Aizawa, K.; Chikazawa, Y.; Ara, K.; Yui, M.; Uemoto, Y.; Kurokawa, M.; Hiramatsu, T. Development of Under Sodium Viewer for next generation sodium-cooled fast reactors. In Proceedings of the International Conference on Fast Reactors and Related Fuel Cycles: Next Generation Nuclear Systems for Sustainable Development (FR-17), Yekaterinburg, Russia, 26–29 June 2017.

12. Chien, H.-T.; Elmer, T.; Engel, D.M.; Lawrence, W.P. Development and Demonstration of Ultrasonic Under-Sodium Viewing System for SFRs. In Proceedings of the International Conference on Fast Reactors and Related Fuel Cycles: Next Generation Nuclear Systems for Sustainable Development (FR-17), Yekaterinburg, Russia, 26–29 June 2017; 2017.

13. Le Bourdais, F.; Le Pollès, T.; Baqué, F. Liquid sodium testing of in-house phased array EMAT transducer for L-wave applications. In Proceedings of the 2015 the 4th International Conference on Advancements in Nuclear Instrumentation Measurement Methods and their Applications (ANIMMA), Lisbon, Portugal, 20–24 April 2015.

14. Cadot, C.; Saillant, J.-F.; Dulmet, B. Method for acoustic characterization of materials in temperature. In Proceedings of the 19th World Conference on Non-Destructive Testing 2016, Munich, Germany, 13–17 June 2016.

15. Saillant, J.F.; Martin, O.; Charrier, S.; Baqué, F.; Sibilo, J. Ultrasonic transducers for Sodium cooled reactors. In Proceedings of the 10th the International Conference on NDE in Relation to Structural Integrity for Nuclear and Pressurized Components, Cannes, France, 1–3 October 2013.

16. Saillant, J.F.; Marlier, R.; Baqué, F. First results of Non-destructive testing under liquid sodium at 200 °C. In Proceedings of the 2016 the IEEE International Ultrasonics Symposium (IUS), Tours, France, 18–21 September 2016.

17. Krautkramer, J.; Krautkramer, H. *Ultrasonic Testing of Materials*, 4th ed.; Springer: Berlin/Heidelberg, Germany, 1990; pp. 528–550.

18. Ancrenaz, P.; Brau, E.; Deydier, S.; Truchetti, L.; Ruos, B. RPV nozzle inner radius (NIR) inspection with a 5 axis robot: Development, qualification and implementation on French reactors. In Proceedings of the 10th International Conference on NDE in Relation to Structural Integrity for Nuclear and Pressurized Components (JRC-NDE 2013), Cannes, France, 1–3 October 2013.

19. Pucci, L.; Raillon, R.; Taupin, L.; Baqué, F. Design of a phased array EMAT for inspection applications in liquid sodium. *Sensors* **2019**. submitted.

20. Le Jeune, L.; Raillon, R.; Toullelan, G.; Baqué, F.; Taupin, L. 2D ultrasonic antenna system for imaging in liquid sodium, sensors 2019. *Sensors* **2019**, submitted.

Review

State-of-the-Art and Practical Guide to Ultrasonic Transducers for Harsh Environments Including Temperatures above 2120 °F (1000 °C) and Neutron Flux above 10^{13} n/cm^2

Bernhard R. Tittmann [1,*], **Caio F.G. Batista** [2], **Yamankumar P. Trivedi** [1], **Clifford J. Lissenden III** [1] **and Brian T. Reinhardt** [3]

1   Department of Engineering Science and Mechanics, Penn State University, University Park, PA 16802, USA; YPT5002@psu.edu (Y.P.T.); cjl9@psu.edu (C.J.L.III)
2   Olympus Corp., State College, PA 16801, USA; caio.batista@outlook.com
3   Applied Research Laboratory, Penn State University, University Park, PA 16802, USA; nairbt85@gmail.com
*   Correspondence: brt4@psu.edu

Received: 2 September 2019; Accepted: 20 October 2019; Published: 1 November 2019

Abstract: In field applications currently used for health monitoring and nondestructive testing, ultrasonic transducers primarily employ PZT5-H as the piezoelectric element for ultrasound transmission and detection. This material has a Curie–Weiss temperature that limits its use to about 210 °C. Some industrial applications require much higher temperatures, i.e., 1000–1200 °C and possible nuclear radiation up to 10^{20} n/cm^2 when performance is required in a reactor environment. The goal of this paper is the survey and review of piezoelectric elements for use in harsh environments for the ultimate purpose for structural health monitoring (SHM), non-destructive evaluation (NDE) and material characterization (NDMC). The survey comprises the following categories: 1. High-temperature applications with single crystals, thick-film ceramics, and composite ceramics, 2. Radiation-tolerant materials, and 3. Spray-on transducers for harsh-environment applications. In each category the known characteristics are listed, and examples are given of performance in harsh environments. Highlighting some examples, the performance of single-crystal lithium niobate wafers is demonstrated up to 1100 °C. The wafers with the C-direction normal to the wafer plane were mounted on steel cylinders with high-temperature Sauereisen and silver paste wire mountings and tested in air. In another example, the practical use in harsh radiation environments aluminum nitride (AlN) was found to be a good candidate operating well in two different nuclear reactors. The radiation hardness of AlN was evident from the unaltered piezoelectric coefficient after a fast and thermal neutron exposure in a nuclear reactor core (thermal flux = 2.12×10^{13} ncm^{-2}; fast flux 2 (>1.0 MeV) = 4.05×10^{13} ncm^{-2}; gamma dose rate: 1×10^9 r/h; temperature: 400–500 °C). Additionally, some of the high-temperature transducers are shown to be capable of mounting without requiring coupling material. Pulse-echo signal amplitudes (peak-to-peak) for the first two reflections as a function of the temperature for lithium niobate thick-film, spray-on transducers were observed to temperatures of about 900 °C. Guided-wave send-and-receive operation in the 2–4 MHz range was demonstrated on 2–3 mm thick Aluminum (6061) structures for possible field deployable applications where standard ultrasonic coupling media do not survive because of the harsh environment. This approach would benefit steam generators and steam pipes where temperatures are above 210 °C. In summary, there are several promising approaches to ultrasonic transducers for harsh environments and this paper presents a survey based on literature searches and in-house laboratory observations.

Keywords: piezoelectric; high-temperature ultrasonic testing; radiation resistance; field-deployable sensor; guided-wave send–receive; spray-on transducers; piezocomposites

1. Introduction

Currently, ultrasonic non-destructive evaluation (NDE) is employed periodically on passive high temperature components, but continuous online monitoring has not been widely implemented. The need for continuous online monitoring is becoming more important with the need for high-temperature infrastructure license extension. Additionally, ultrasound is a highly attractive NDE methodology given that it allows for inspection in optically opaque materials, such as liquid-metal coolants, steam generator piping, and heat exchanger pipes. Further applications may be found in materials research reactors where ultrasonic NDE can be used for in situ analysis of radiation effects on novel radiation-hard materials currently being developed.

During the past decades there has been significant interest and therefore research into the problem of ultrasonic transducers for harsh environments [1–82]. The aim of this paper is to give an overview, review, and survey of piezoelectric materials for possible harsh-environment applications. The survey is conveniently divided into several categories: single crystals, piezoelectric ceramics, composite ceramics, and radiation-resistant materials and the new category of brush-on transducers. The survey starts with several relatively well-known high-temperature piezoelectric materials summarized in Table 1 for comparison. Listed also are the Curie–Weiss temperatures, which are useful in that they provide a limit to the temperature to which a material can exhibit piezoelectricity. Furthermore, the conventional PZT 5H is also listed, which is the commonly used piezoelectric in commercial applications.

Table 1. Some well-known piezoelectrics [1–9].

Piezoelectric Material	Curie–Weiss Temperature (°C)
PZT-5H	210
Keramos lead metaniobate	400
Bismuth titanate	685
Lithium niobate	1000

2. Transducers for High Temperature Applications

2.1. Single-Crystal Wafers

In the category of the single crystals, both maximum temperature and long-term in situ operation were investigated in a comparison study. These is the well-known lithium niobate (LiNbO$_3$), and then two relatively recently developed materials [3]: aluminum nitride (AlN) and YCOB [YCa$_4$O(BO$_3$)$_3$]. As shown in Figure 1, the pulse-echo amplitude of LiNbO$_3$ is stable until about 1000 °C [1]. Figure 2 shows the pulse-echo amplitude response for a single-crystal wafer of aluminum nitride at two temperatures, 25 and 750 °C, showing only somewhat lower amplitudes at the higher temperature [2]. Figure 3 shows the ultrasonic high-temperature performance of single-crystal AlN wafer coupled to a steel cylinder with acceptable performance to about 950 °C [2].

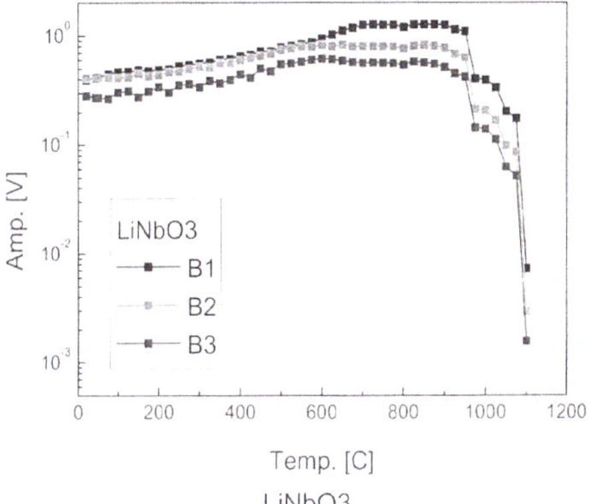

Figure 1. Temperature profiles for pulse-echo amplitude of lithium niobate (LiNbO$_3$) single crystal bonded to steel. B1, B2, and B3 describe three successive runs [1].

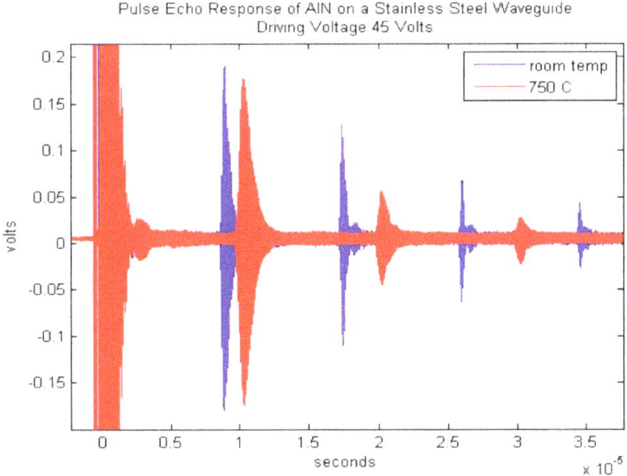

Figure 2. Pulse-echo amplitude response for a single-crystal wafer of aluminum nitride at two temperatures, 25 and 750 °C, showing only somewhat lower amplitudes at the higher temperature [2].

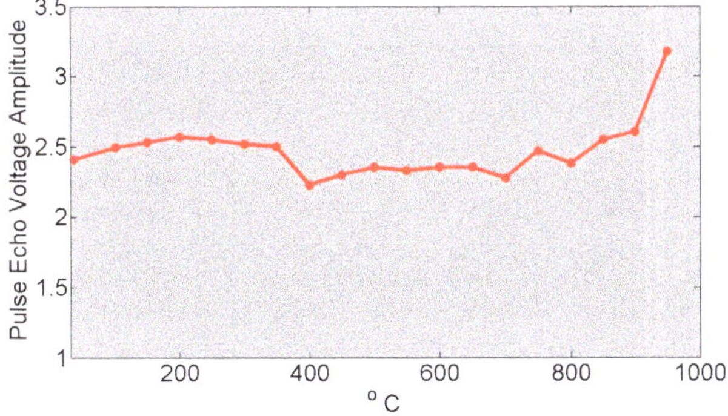

Figure 3. Ultrasonic high-temperature performance of single-crystal AlN wafer on steel cylinder showing acceptable performance to about 950 °C [2].

Shown in Figure 4 are three consecutive runs over a measurement time of 14 h [3]. As can be seen, all three materials exhibited stability in ultrasonic performance through heat treatment of 950 °C for 24 h and 1000 °C for 48 h. This "cook-and-look" testing revealed significant changes in the dielectric properties and only small changes in the ultrasonic performance of lithium niobate. Dielectric changes of the observed magnitude would be expected to have a noticeable effect on the ultrasonic performance. However, the heat treatments were not equivalent during the dielectric and ultrasonic testing. It is quite likely that the longer heat treatment caused a more pronounced change in the dielectric properties of the lithium niobate [3].

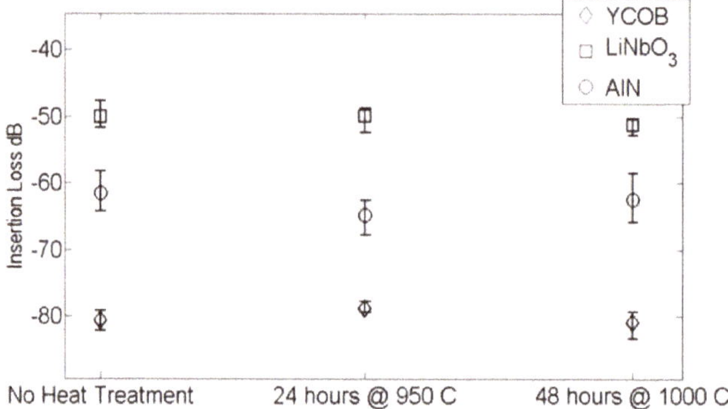

Figure 4. Comparison of heat treatment results for lithium niobate (LiNbO$_3$), aluminum nitride (AlN), and YCOB [YCa$_4$O(BO$_3$)$_3$] [3].

The YCOB on the other hand exhibited a much less pronounced change in dielectric properties after heat treatment. It is expected that YCOB is more stable at high temperatures than LiNbO$_3$ which is known to deplete its oxygen particularly at low oxygen partial pressure [3].

Material selection is based primarily on combining Curie temperatures (T$_c$) and coupling coefficients (e.g., d$_{33}$) of the constituents to achieve the desired overall piezoceramic properties. To maintain an in-field transducer at high signal-to-noise (SNR), the piezoelectric transducer material should have both a large coupling coefficient and a T$_c$ exceeding the transducer's operating temperature [4–10]. Micromechanical modeling enables prediction of overall properties based on the properties of the constituents. Figure 5 presents a graph showing the electromechanical coefficient, d$_{33}$, as a function of the T$_c$ for a selection of piezoelectric materials, consisting of single crystals, polycrystals, textured crystals, and films [10].

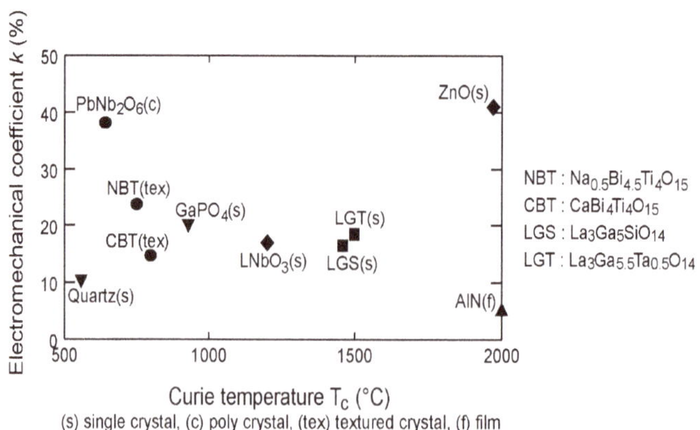

Figure 5. Electromechanical coefficient versus Curie temperature T$_c$ [10].

2.2. Thick-Film Ceramics

In the category of thick-film ceramic sample preparation, poling, acoustic data, high-temperature tests, and the effect of protective aluminum oxide layer on both poling and temperature performance were studied. Bismuth titanate thick-film transducers performed well up to 600 °C. Currently, tests are ongoing with thick-film transducers deposited on pipes and simulated casings for NDE with guided waves generated by both flat and curved arrays. Recently developed piezoelectric materials with high Curie temperatures are listed in Table 2.

Table 2. High-temperature piezoelectric ceramics.

Piezoelectric Material	Curie Temperature (°C)
Praseodymium titanate	>1550 [11,12]
Lanthanum titanate	1461 [13,16]
Neodymium titanate	1482 [13,17]
Strontium niobate	1327 [14]
Calcium niobate	>1525 [15]

Conventional piston-type transducers that send and receive ultrasonic waves typically use lead zirconate titanate for the active element and have backing and matching layers. In addition, they are usually coupled to the substrate through gel or adhesive. Harsh environments limit the types of couplants that can be used, and curved surfaces present additional challenges. In contrast, spray-on transducers are bonded directly to the substrate, precluding the need for couplants. Spraying transducers onto curved surfaces is not substantially different from doing so on flat surfaces. No matching or backing layers are used in this work, but they could be used if deemed necessary. One advantage that spray-on transducers provide is the ability to design the transducer material for a specific operating temperature by mixing powders into a sol gel to create a composite (or alloy).

2.3. Composite Ceramics

The biggest difference between piezoelectric materials used in conventional transducers and spray-on piezoelectric transducers is density/porosity. Pressure is an integral part of forming fully dense piezoceramics, and it is not part of spray-on processing. Thus, spray-on transducers have porosity that affects their properties. On the positive side, it also provides strain tolerance to the piezoceramic, which is bonded to a metal substrate that is subject to temperature changes. The pioneers of spray-on piezoelectric transducer technology are Barrow and Kobayashi. Barrow et al. [18,19] added powder to sol gel to form piezoelectric films thicker than 1–2 μm using a spin coating methodology. Kobayashi et al. [20–24] then adapted the powder/sol–gel technique using a spray gun to deposit films on metal substrates. Searfass et al. [25–28] have provided technological advancements on the sol–gel composition, fabrication, characterization, and high-temperature ultrasonic testing for such spray-on transducers. As examples, Figures 6 and 7 show that PZT/$Bi_4Ti_3O_{12}$ and $Bi_4Ti_3O_{12}$/$LiNbO_3$ composite transducers mounted on steel cylinders functioned well in pulse-echo mode until 675 and 1000 °C, respectively.

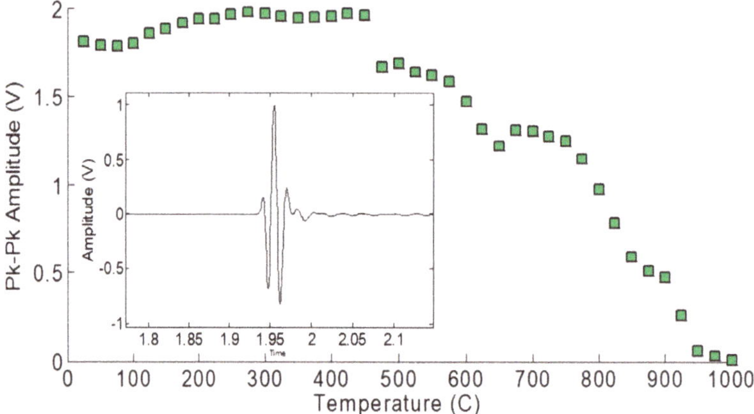

Figure 6. Temperature dependence of pulse-echo amplitude for PZT/Bi$_4$Ti$_3$O$_{12}$ piezocomposite spray-on transducers deposited on steel cylinders [29].

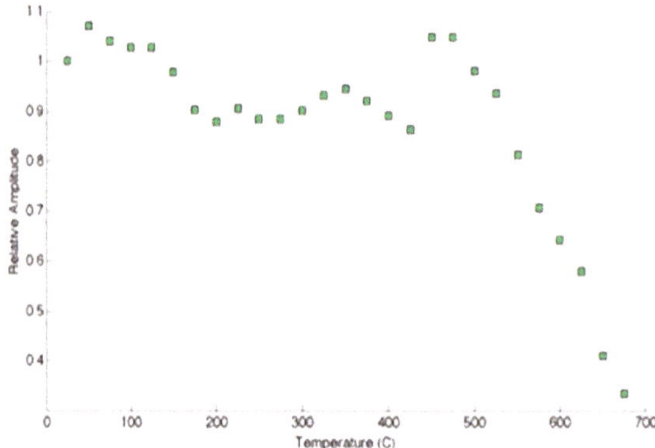

Figure 7. Temperature dependence of pulse-echo amplitude for Bi$_4$Ti$_3$O$_{12}$/LiNbO$_3$ piezocomposite spray-on transducer [29].

The PZT/Bi$_4$Ti$_3$O$_{12}$ transducer's efficiency decreased when operating in pulse-echo mode, but a discernable signal was still observed as low as 500 kHz. The thickness of this transducer was still relatively thin, especially for low-frequency operation. The broadband nature of this transducer was very evident in its testing in that it had a center frequency around 2.75 MHz but could still operate effectively well below 1 MHz.

The Bi$_4$Ti$_3$O$_{12}$/LiNbO$_3$ transducer was also tested for low-frequency operation, but it was considerably less efficient. The signal effectively disappeared at frequencies much below 1 MHz. This again shows the great advantage of the use of the PZT/Bi$_4$Ti$_3$O$_{12}$ composite. The PZT/Bi$_4$Ti$_3$O$_{12}$ has a much greater signal amplitude and is more broadband allowing it to operate at low frequencies and produce viable waveforms. Thicker PZT/Bi$_4$Ti$_3$O$_{12}$ transducers may further enhance their operation at low frequencies. Both the signal amplitude and signal-to-noise ratio can be increased along with better operation in the pulse-echo mode [25–28].

3. Piezoelectric Materials for Radiation Environment

Ultrasonic measurements have a long and successful history of use for material characterization, including detection and characterization of degradation and damage, measurement of various physical parameters used for process control, such as temperature and fluid flow rate, and in nondestructive evaluation (NDE) [31]. However, application of ultrasonic sensors in nuclear reactors has been limited to low neutron flux environments. The development of ultrasonic tools to perform different in-pile measurements requires a fundamental understanding of the behavior of ultrasonic-transducer materials in these high neutron flux environments. Irradiation studies of ultrasonic transducers have been described in the literature but are generally at lower flux/fluences than what might be seen in U.S. nuclear reactors. The Pennsylvania State University (PSU) lead an effort that was selected by the Advanced Test Reactor National Scientific User Facility (ATR-NSUF) for an irradiation of ultrasonic transducers in the Massachusetts Institute of Technology Nuclear Research Reactor (MITR) [32,33]. This test was an instrumented-lead test, allowing real-time signals to be received from five ultrasonic transducers including three piezoelectric transducers two of which were single-crystal wafers of aluminum nitride. The irradiation began on February 20th, 2014 and was scheduled to run for a period of 18 months or until all the sensors have ceased to operate. Recent results are presented and discussed in detail in References [32,33]. In searching for candidate materials for use in harsh environments, the most straightforward down selection parameter seems to be the transition temperature, which provides an upper limit on the operating range of the piezoelectric material. In fact, a higher Curie temperature has been found to correlate with increased radiation tolerance and the primary effect of radiation damage in piezoelectric materials appears to be depolarization [21]. A table of candidate materials for longitudinal wave generation is provided below in Table 3; however, this is only the first step. The final column in Table 3 is of substantial importance as it has been found that crystal structure plays a significant role in radiation tolerance of ceramics [33–74].

Table 3. Piezoelectric materials [33].

Material	Transition Temperature °C	Transition Type	Structure
AlN	2826	Melt	Wurtzite [6]
Bi_3TiNbO_9	909	Curie	Perovskite layered [32]
$LiNbO_3$	~1200	Curie	Perovskite [21]
$Sr_2Nb_2O_{73}$	1342	Curie	Perovskite layered [32]
$La_2Ti_2O_7$	1500	Curie	Perovskite layered [32]
$GaPO_4$	970	α-β	SiO_2 homeotype [32]
$ReCa_4(BO_3)_3$, *Re as Rare Earth element*	>1500	Melt	Oxyborate homeotype [32]
ZnO	1975	Melt	Wurtzite [33]

For the radiation effects, the discussion focuses on the case of AlN, which is not a ferroelectric and has a transition temperature of 2865 °C (melting temperature). We also consider the case in which the bulk of the crystal is kept below any transition temperature. In this scenario, during irradiation four primary forms of damage are anticipated in a piezoelectric material:

(1) depoling via thermal spike processes,
(2) morphization/metamictization due to displacement spikes or high concentration of point defects,
(3) increase in point defect concentration, and
(4) development of defect aggregates.

Here, only the two most likely damage mechanisms are summarized, namely thermal spikes and displacement spikes [33]. Additionally, transmutation products are considered, as these in fact induce both thermal spikes and displacement spikes in some cases. To summarize, the considerations lead to the conclusion that AlN is resistant to amorphization. Moreover, the very high transition temperature renders the material immune to thermal spike damage. It is also clear that the transmutation reaction, $^{14}N(n,p)^{14}C$, generates only a fraction of a dpa at 10^{21} n/cm^2 and insignificant doping.

A single-crystal AlN element (4.8 mm in diameter and 0.45 mm thick) resonant at 13.4 MHz, was coupled to an aluminum cylinder via mechanical pressure. Aluminum foil was used as an acoustic coupler between the aluminum cylinder and the AlN element, allowing for strong, clear A-scan data to be obtained. The AlN element was loaded, on the side opposite the aluminum cylinder, with a sintered carbon/carbon composite to reduce ringing and improve the signal clarity. The test fixture is illustrated in Figure 8.

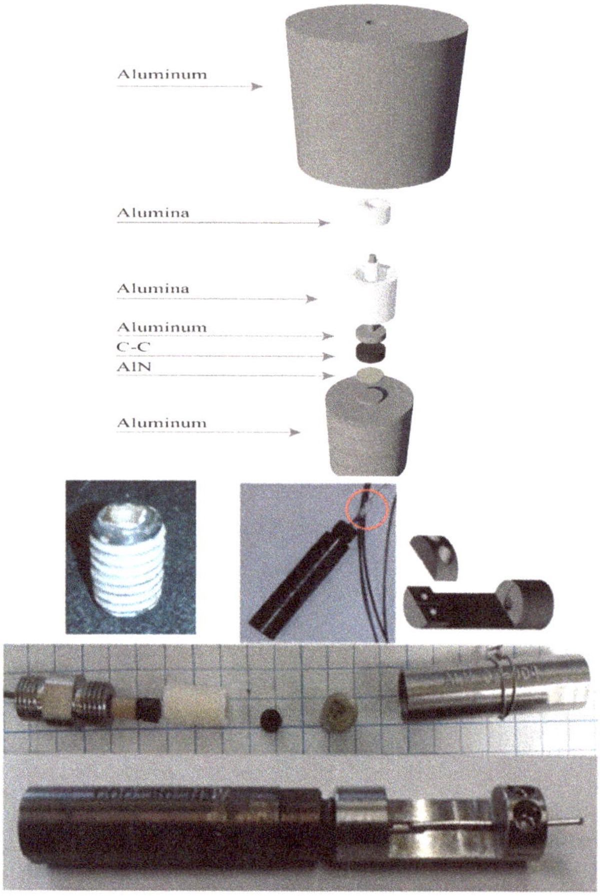

Figure 8. Photos of the fixture inserted into the Massachusetts Institute of Technology (MIT) reactor [31].

The aluminum cylinder acted as the lower electrical contact and the plunger provided the upper electrical contact. The setup was connected to a radiation-hard 50 ohm coaxial cable. This radiation-hard cable consisted of an aluminum conduit sleeve over fused quartz dielectric tubing with an aluminum inner conductor. The cylinder/piezo setup was placed in the core of the Penn State TRIGA reactor and irradiated to a fast and thermal neutron fluence of 1.85×10^{18} n/cm^2 and 5.8×10^{18} n/cm^2, respectively, and a gamma dose of 26.8 MGy. Throughout the irradiation the A-scan data were recorded with impedance measurements interspersed.

A similar fixture was built and inserted into the reactor at the Massachusetts Institute of Technology (MITR) for the ATR-NSUF tests. Table 4 gives the MIT Research Reactor Environment. Figure 8 shows a photo of the fixture before being inserted into the MITR.

Table 4. MIT research reactor environment.

The Massachusetts Institute of Technology Reactor is characterized by the following features:
Total flux = 1.89×10^{14} n/cm^2 Thermal flux (<0.4 eV) = 2.12×10^{13} n/cm^2 Epi-thermal flux (0.4 eV–0.1 MeV) = 8.03×10^{13} n/cm^2 Fast flux 1 (>0.1 MeV) = 8.78×10^{13} n/cm^2 Fast flux 2 (>1.0 MeV) = 4.05×10^{13} n/cm^2 Gamma dose rate: 1×10^9 r/h   Temperature: 400–500 °C

3.1. Temperature Tolerance

Prior high-temperature experiments with AlN [2,3] may lead one to suspect that crystalline defects can degrade the high-temperature transduction of AlN. Considering that radiation causes displacement damage and transmutation doping, one may wonder how the irradiated AlN would fare at high temperatures. To answer this call the irradiated crystal, having negligible activity after cooling for a few weeks, was tested up to 500 °C. Figure 9 shows the relative pulse-echo amplitude measured as a function of temperature. Some of the waveforms are provided in Figure 10. Additionally, d_{33} was measured prior to and after irradiation and found to be 5.5 pC/N, which is unchanged from the pristine value. Further, subjecting the irradiated AlN crystal to temperatures of 950 °C for 72 h caused no change in the performance of the AlN crystal [32–34].

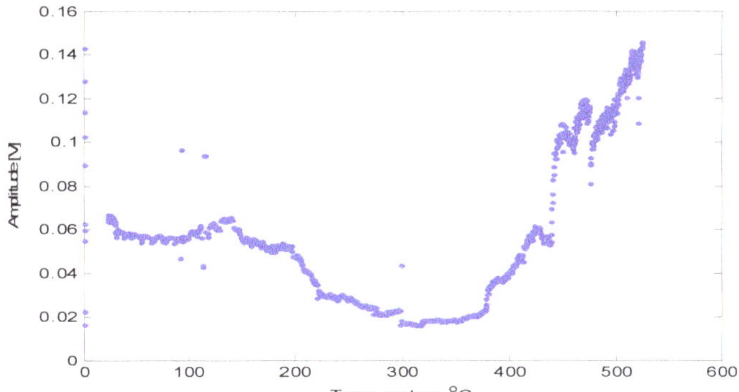

Figure 9. Relative pulse-echo amplitude measured as function of temperature for AlN sample. Note the increase as temperature is raised above 400 °C [31].

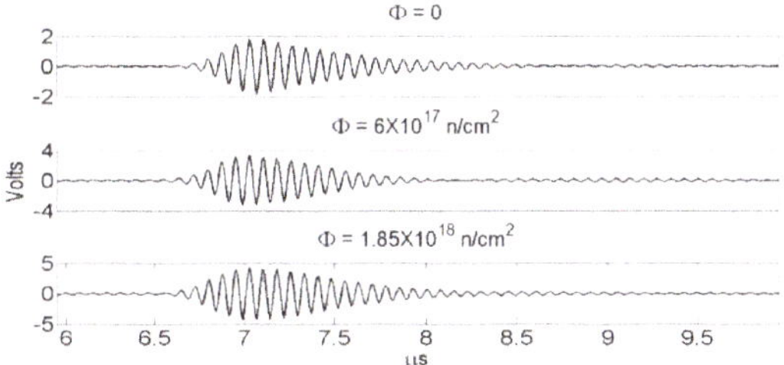

Figure 10. A-Scans obtained from AlN TRIGA reactor core, Φ is the fast neutron fluence [31].

3.2. Radiation Tolerance

The A-scan data, illustrated in Figure 10, were recorded and analyzed in terms of the echo amplitude, which are presented in Figure 11. The amplitude over the course of irradiation remains nearly constant and indicates the radiation hardness of the AlN and the test fixture. In Figure 11 the black dots represent the data from the tests in the Penn State TRIGA reactor, whereas the red points represent the data from the tests in the MITR.

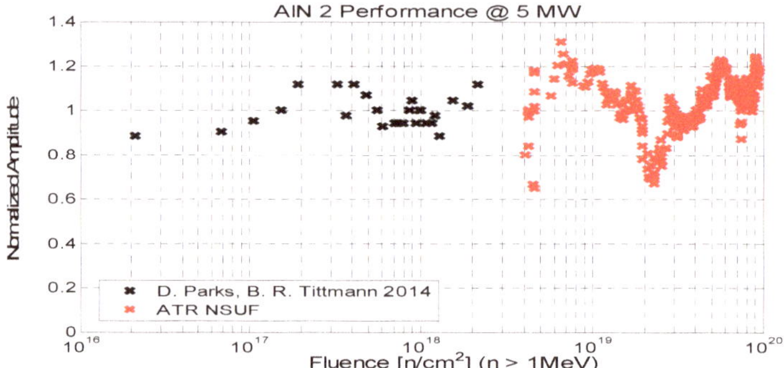

Figure 11. Normalized amplitude of pulse-echo signal showing the results of both the Pennsylvania State University (PSU) TRIGA reactor and the MITR (Massachusetts Institute of Technology Nuclear Research Reactor) measurements. Note that the excursions to low amplitudes are the results of reactor scrams [31].

For practical use in harsh radiation environments, the selection criteria for piezoelectric materials for NDE and material characterization were summarized. Using these criteria piezoelectric aluminum nitride was shown to be a viable candidate. The results of tests on an aluminum-nitride-based transducer operating in two nuclear reactors were presented. The tolerance of single-crystal piezoelectric aluminum nitride after fluences of up to 10^{20} n/cm^2 is examined. The radiation hardness of AlN is most evident from the unaltered piezoelectric coefficient d_{33}, which measured 5.5 pC/N after a fast and thermal neutron exposure in a nuclear reactor core for over 120 MWh in agreement with the published literature value. The results offer potential for improving reactor safety and furthering the understanding of radiation effects on materials by enabling structural health monitoring and NDE in spite of the high levels of radiation and high temperatures known to destroy typical commercial ultrasonic transducers.

4. Spray-On Transducers for Harsh Environment Applications

Damage detection in the power industry is always vying for optimized and cheaper techniques. Most components in the energy sector utilize metallic structure, whether it is for power generation, storage, transportation, or waste management. Many components operate at a high temperature adding further challenges for their health monitoring. Given that commercial transducers rated for elevated temperatures are limited and expensive, the use of spray-on film transducers for such purposes has been researched while keeping the fabrication simple enough for anyone to create them.

Bismuth titanate (Bi$_4$Ti$_3$O$_{12}$) is an excellent piezoelectric which has a T$_c$ of 670 °C and a safe operating level until about 500 °C, considerably higher than PZT. Unlike the preceding sol–gel method [18–26], this fabrication process involves a lithium-silicate-based inorganic binder and water to mix with the Bi$_4$Ti$_3$O$_{12}$ powder. The following steps are optimized for best results.

1. Select the powder (BIT or lithium niobate/barium titanate) and mix with Ceramabind 830 to achieve a 1:0.2:0.8 ratio (powder–binder–water by weight ratio); A plastic stirrer was used to rigorously mix the powder and binder, but it could be mixed with a ultrasonic horn.
2. Create the solution by combining the mixed powder/binder with distilled water at the specified concentration in a 15 mL glass vial.
3. Prepare the substrate by roughening the surface with a fine-grit sandpaper, and then clean it with isopropyl alcohol.

4. Spray the slurry onto the substrate with an air gun (Goplus Electric Paint Sprayer, 450W High Power HVLP Paint Spray Gun with 3 Spray Patterns, 3 Nozzle Sizes, Adjustable Valve Knob and 900ml Large Detachable Container); The air gun pressure should be 20–22 psi and the nozzle should be approximately 20 cm from the surface. Alternatively, apply slurry with a brush.

5. Dry each layer of the sprayed film in the relatively low-humidity environment (15–20%) of a glove box for at least 15 min to avoid cracking.

6. Repeat steps 4 and 5 to achieve the desired film thickness (preferably thicker than 120 μm). The average thickness of a single spray is 18 μm.

7. Use a thickness gage to measure the average thickness of the film.

8. After the film layers have cured, brush apply a conductive silver paint (SPI Chemicals, Inc., Atlanta, GA, USA) on the portion of the film to become the transducer to a thickness of approximately 30 μm. Each layer takes approximately 15 min to cure in the low-humidity setting, so if there are eight spray repetitions, it will take about 2 h. For films thicker than eight layers, the cure time for a layer may be longer.

9. Once the electrode is applied, heat the sample to 60 °C for a few minutes with a heat gun to allow the electrode to dry. This step is optional as the electrode can air dry in a longer time.

10. Attach a bare nickel chrome wire (supplied by Consolidated) with silver paint to serve as the lead wire as shown in Figure 12.

11. Pole sample at a desired electric field for at least 20 min at ambient temperature.

Figure 12. Bismuth titanate, lithium niobate, and organic bismuth titanate film transducers (left to right). Good adhesion was observed for extended periods of time. One film transducer is still working normally after several years of use [7].

With the initial bulk-wave characterization of the film, it was noted that despite the lower piezoelectric coefficient compared to PZT, the film transducers were able to function at higher temperatures. Another major advantage is the straightforward fabrication procedure and the ability for these films to cure at room temperature. An alternative method to produce these films consisted of using organic compounds instead of the high-temperature inorganic binder. The organic films were also excellent in heat resistance despite having a slightly complicated fabrication procedure compared to the inorganic films. Figure 12 shows samples of $Bi_4Ti_3O_{12}$ (left), $LiNbO_3$ (center), and organic $Bi_4Ti_3O_{12}$ (right) spray-on films were fabricated using the beforementioned procedure and the inorganic method.

Figure 13 shows A-scan pulse-echo measurements of $Bi_4Ti_3O_{12}$ (left), $LiNbO_3$ (center), and organic $Bi_4Ti_3O_{12}$ (right) thick-film transducers deposited on a steel cylinder (7 mm) at 40 dB gain, film thickness ~200 micron, and frequency ~1.5 MHz [7].

The bismuth titanate powder used was 99.99% pure $Bi_4Ti_3O_{12}$ 200 mesh (75 μm particle size) supplied by Lorad Chemical Corporation. Lithium niobate powder was a 99.99% pure $LiNbO_3$ 325 mesh (45 μm particle size) supplied by LTS Research Laboratories, Inc. (Orangeburg, NY 10962, USA). The barium titanate used was 99% pure $BaTiO_3$ 325 mesh (45 μm particle size) supplied by Acros organics (Thermo Fisher Scientific, Branchburg, NJ, USA).

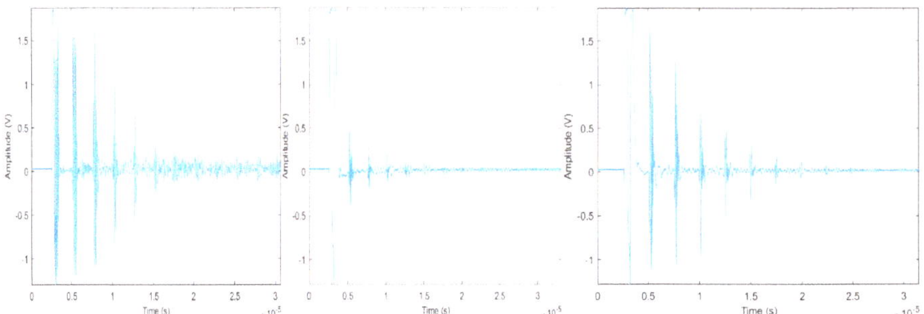

Figure 13. A-scan pulse-echo measurements of $Bi_4Ti_3O_{12}$ (**left**), $LiNbO_3$ (**center**), and organic $Bi_4Ti_3O_{12}$ (**right**) thick-film transducers deposited on steel cylinder (7 mm) at 40 dB gain, film thickness ~200 micron, and frequency ~1.5 MHz [7].

The Curie–Weiss temperatures of $Bi_4Ti_3O_{12}$ and $LiNbO_3$ make them ideal candidates for high-temperature testing. These films were inserted in the tube furnace and peak-to-peak voltage measurements for the first and second reflection from an edge were recorded. The furnace was set to increase the temperature of the films at a rate of about 6 °C/min. The $Bi_4Ti_3O_{12}$ films were tested up to a temperature of 650 °C, whereas the $LiNbO_3$ films were tested to a temperature of 900 °C. The first and second echoes were recorded in terms of the signal amplitude and plotted in relation with the temperature ramp seen here in Figure 14 for the $Bi_4Ti_3O_{12}$ film. Figure 15 shows the results for $LiNbO_3$ sample.

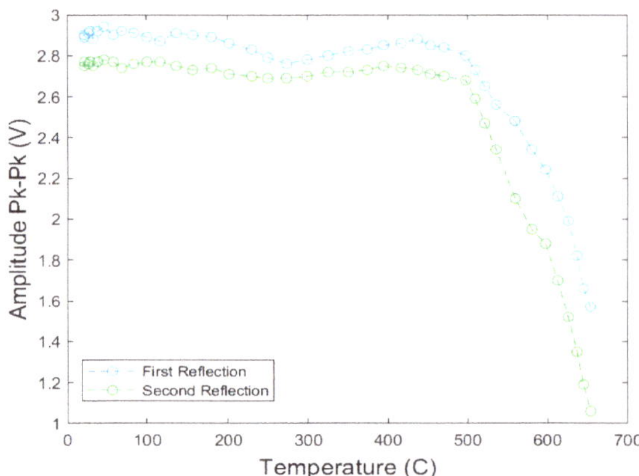

Figure 14. Signal amplitude (peak-to-peak) for the first two reflections as a function of temperature for bismuth titanate thick-film, spray-on transducer [7].

To demonstrate their ability to perform as a guided-wave sensor, the primary Lamb wave modes (A0, S0, A1, S1) were generated using a comb transducer arrangement. Furthermore, 6061 aluminum plates of several different thicknesses ranging from 2 to 4 mm were chosen as the waveguides. Sets of comb transducers were then applied onto the plate a certain distance apart in a through-transmission setup as shown in Figure 16a,b.

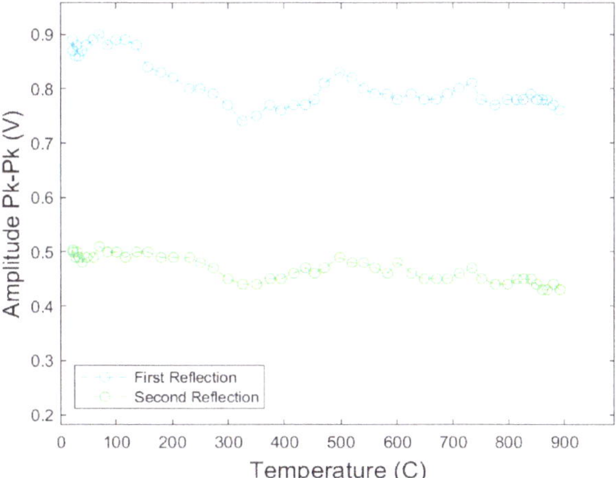

Figure 15. Signal amplitude (peak-to-peak) for the first two reflection as a function of the temperature for lithium niobate thick-film, spray-on transducer [7].

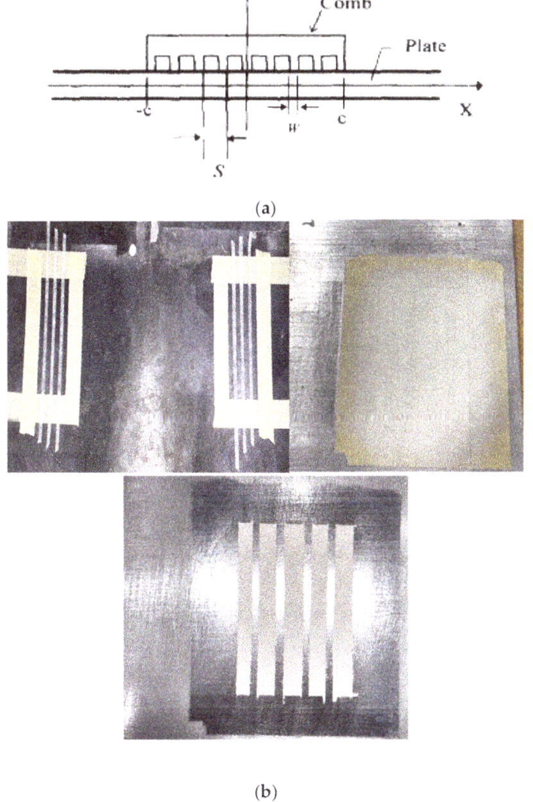

Figure 16. (a) S is the element width, W is the element gap, W + S is the wavelength of the traveling wave [7]; (b) Schematic of the experiment setup for generation of Lamb waves in a 3.2 mm thick 6061 aluminum plate [7].

The number of actuating and receiving elements were altered to give rise to various configurations for better transducer characterization. Calculations solving the thin-plate Lamb wave transcendental equations were performed for the comb elements to be spaced by the same length as the wavelength of the preferential excited mode [79]. See Figure 17 for graphs of the corresponding dispersion curves. A tone-burst of 15 cycles was introduced in the actuator set of the transducers and the A-scan was plotted. According to the excitation parameters and the comb spacing, the S1 mode should be the first to be received as shown in Figure 18. The readings were then recorded through various receiving elements and calculation were performed to compare the experimental group velocity for S1 mode with its theoretical value at that specific frequency.

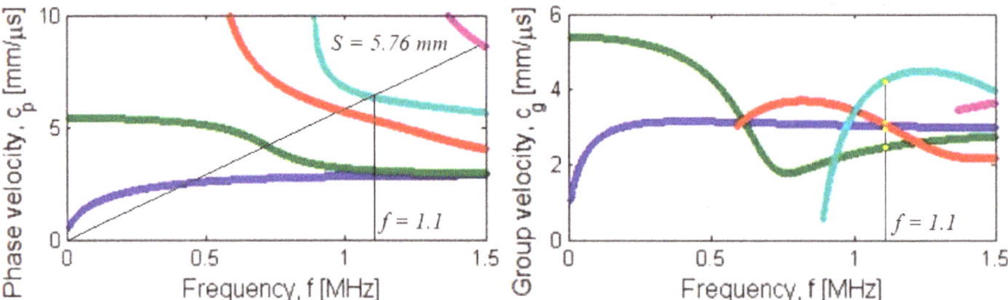

Figure 17. Dispersion curves for a 3.2 mm thick aluminum 6061 plate showing the activation line on the phase velocity and corresponding group velocities at that frequency [82].

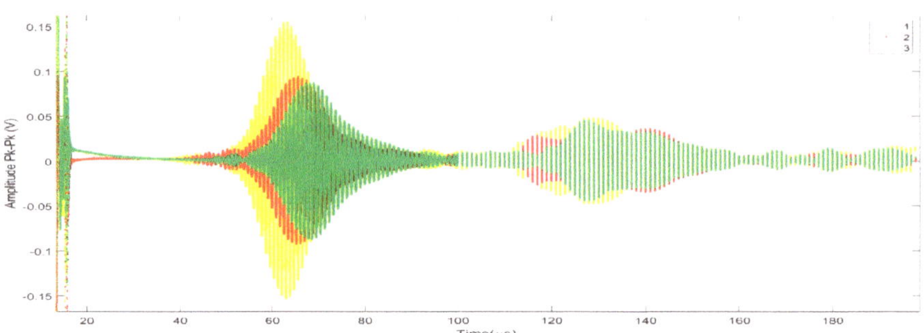

Figure 18. A-scan for one of the tested plates showing the superimposed guided-wave modes from three different elements [7].

Figure 19 presents the frequency spectrum for one of the received waveforms showing both the second and third harmonics. The third harmonic appears relatively strong and useful for further studies. The other A-scan waveforms display similar frequency spectra.

Signal-to-noise ratio (SNR)was calculated by taking the ratio of the root-mean-square (rms) value of the amplitude (peak-to-peak) within the mode window to the rms value of the noise window in decibel units given by this formula.

$$SNR = 20 * log_{10}\left(\frac{V_{pk-to-pk}}{V_{noise}}\right)$$

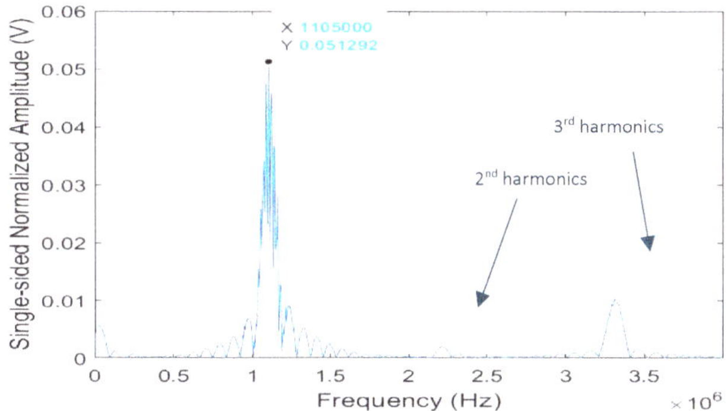

Figure 19. Frequency spectrum for one of the guided modes. The presence of harmonics is also indicated [7].

The rms value is calculated by multiplying $1/(2\sqrt{2})$ to the peak-to-peak voltage. The signal-to-noise ratio along with the signal strength for the three films is shown in Table 5.

Table 5. Signal-to-noise calculations for the inorganic and organic films.

Film	$Bi_4Ti_3O_{12}$	$LiNbO_3$	Organic $Bi_4Ti_3O_{12}$
Signal window (µs)	5.071–5.739	5.071–5.554	5.030–5.635
Noise window (µs)	5.756–7.510	5.615–7.490	5.675–7.450
Signal strength pk-pk (V)	2.895	0.868	2.815
Noise strength pk-pk (V)	0.267	0.156	0.284
Signal rms (V)	1.024	0.307	0.996
Noise rms (V)	0.094	0.055	0.100
Signal strength (dB)	9.321	−1.232	8.990
Noise strength (dB)	−11.484	−16.157	−10.939
Signal-to-noise ratio (dB)	20.716	14.926	19.929

The inorganic films are easy to produce and extremely inexpensive, as the films use very small quantities. A batch of 5 g powder with another 5 g of the solvents is enough to coat at least two of the stainless cubes used for the pulse-echo measurements. For spraying films on bigger areas, the slurry quantity would also have to be increased, but compared to the cost of commercial sensors these transducers are affordable. The respective SNR of 20.72, 14.93, and 19.93 dB for the $Bi_4Ti_3O_{12}$, $LiNbO_3$, and organic $Bi_4Ti_3O_{12}$ films is comparable to commercial transducers. According to the study done by Kobayashi, the films on the planar surfaces yielded an SNR of 16 dB and a center frequency of 3.6 MHz, where the $Bi_4Ti_3O_{12}$ was doped with PZT. Here, without a strong piezoelectric material such as PZT, the signal strength is quite strong. The center frequencies for the $Bi_4Ti_3O_{12}$, $LiNbO_3$, and organic $Bi_4Ti_3O_{12}$ films were 1.4, 1.22, and 1.42 MHz, respectively. With the premise of a working transducer, the films were then subjected to temperature testing. From the capacitance, other parameters to characterize the films could be calculated such as the dielectric constant, dielectric loss, amount of charge in the film, and even piezoelectric constant after some analysis. The higher the dielectric constant, more charge could be held by it and used as electric potential. Poling these films requires patience. Through some further study, it was found that an aluminum oxide protective layer on the top of the piezo film was useful for serving as an electrical blanket, which prevented the charges from jumping and short circuiting the film. With the alumina layer, the films could be poled at voltages only a few hundred volts below the coercive field.

5. Conclusions

This Invited Special Issue contribution addresses the current state-of-the-art and offers a practical guide to ultrasonic transducers for harsh environments including temperatures above 2120 °F (1000 °C) and neutron flux above 10^{13} n/cm^2.

In field applications currently used for health monitoring and nondestructive testing, ultrasonic transducers primarily employ PZT5-H as the piezoelectric element for ultrasound transmission and detection. This material has a Curie–Weiss temperature which limits its use to about 210 °C. Some industrial applications require much higher temperatures, i.e., 1000–1200 °C and possible nuclear radiation up to 10^{20} n/cm^2 when performance is required in a reactor environment.

The goal of this paper is the survey and review of piezoelectric elements for use in harsh environments for the ultimate purpose for Structural Health Monitoring (SHM), Non-destructive Evaluation (NDE) and material characterization (NDMC). The survey comprised the following categories: 1. High temperature applications with single crystals, thick film ceramics, and composite ceramics, 2. Radiation tolerant materials, and 3. Spray-on transducers for harsh environment applications. In each category the known characteristics are listed, and examples are given of performance in harsh environments.

In summary we have presented a survey of piezoelectric materials capable of operation at higher temperatures and possible nuclear radiation. This survey tries to both gather information and summarize it. The findings indicate that PZT/$Bi_4Ti_3O_{12}$ and $Bi_4Ti_3O_{12}$/$LiNbO_3$ composite transducers functioned in pulse-echo mode until 675 and 1000 °C, respectively. Recent interest in the radiation endurance of piezoelectric ultrasonic transducers has stimulated a search for appropriate materials. Some applications may be found in materials research reactors where ultrasonic NDE can be used for in situ analysis of radiation effects on novel radiation-hard materials currently being developed. This paper presents a survey of piezoelectric materials for possible harsh-environment applications. Moreover, our experiments in a nuclear reactor for one of the materials, AlN, demonstrated an example of possible resistance to radiation. Unfortunately, AlN is not a very efficient producer of ultrasonic waves. Therefore, one of the future goals is to come up with transducers with higher efficiency that are tolerant of radiation. We also showed that some of the high-temperature transducers could be mounted on a target without requiring coupling material. Guided-wave send-and-receive was demonstrated on planar and pipe structures for possible field deployable applications. Interesting and possibly relevant research applications with brush-on transducers are going on at the University of Montpellier for fission gas characterization [83,84].

Author Contributions: Conceptualization: B.R.T. was invited to contribute to a Special Issue in *Sensors* on the topic of "Ultrasonic Transducers for Harsh Environments". The article was first conceived by B.R.T. and then developed with contributions from all the authors. C.F.G.B. was key to developing the topic of ceramic composites; Whereas Y.P.T. was instrumental in the topic of brush-on/spray-on transducers, which is part of his M.S. Thesis at Penn State University under the supervision of C.J.L.III. Writing: The Original Draft was developed by B.R.T.; Writing—Review & Editing, was done by C.F.G.B. and B.T.R.; The final editing was done by B.R.T.

Funding: Was in part by US Department of Energy, Office of Nuclear Energy under DOE Idaho Operations Office Contract DE-AC07-051D14517.

Conflicts of Interest: The authors declare no conflict of interest.

References

1. Baba, A.; Searfass, C.T.; Tittmann, B.R. High temperature ultrasonic transducer up to 1000°C using lithium niobate single crystal. *AIP Appl. Phys. Lett.* **2010**, *97*, 232901. [CrossRef]
2. Parks, D.A.; Kropf, M.M.; Tittmann, B.R. Aluminum Nitride as a High Temperature Transducer. In Proceedings of the 36th Annual Review of Progress in Quantitative Nondestructive Evaluation, Kingston, RI, USA, 26–31 July 2009.

3.   Parks, D.A.; Zhang, S.; Tittmann, B.R. High-Temperature (>500 °C) Ultrasonic Transducers: An Experimental Comparison Among Three Candidate Piezoelectric Materials. *IEEE Trans. Ultrason. Ferroelectr. Freq. Control* **2013**, *60*, 1010–1015. [CrossRef] [PubMed]

4.   Searfass, C. Characterization of Bismuth Titanate Thick Films Fabricated Using a Spray-On Technique for High Temperature Ultrasonic Non-Destructive Evaluation. Ph.D. Thesis, The Pennsylvania State University, Stecker Ridge, PA, USA, 2012.

5.   Pheil, C.S. Fabrication and Testing of High Temperature Ultrasonic Transducers. Bachelor's Thesis, The Pennsylvania State University, Stecker Ridge, PA, USA, 2012.

6.   Reinhardt, B. Nonlinear Ultrasonic Measurements in Nuclear Reactor Environments. Ph.D. Thesis, The Pennsylvania State University, Stecker Ridge, PA, USA, 2016.

7.   Trivedi, Y. Field-Deployable Guided Wave Transducers for High-Temperature Applications. Master's Thesis, The Pennsylvania State University, Stecker Ridge, PA, USA, 2016.

8.   Ledford, K.R. Practical Sprayed-on Transducer Composites for High Temperature Applications. Master's Thesis, The Pennsylvania State University, Stecker Ridge, PA, USA, 2015.

9.   Xu, J.L. Practical Ultrasonic Transducers for High Temperature Applications using Bismuth Titanate and Ceramabind. Master's Thesis, The Pennsylvania State University, Stecker Ridge, PA, USA, 2017.

10. MackertichSengerdy, G. Aluminum Nitride Transducer Design for Harsh Environment. Master's Thesis, The Pennsylvania State University, Stecker Ridge, PA USA, 2018.

11. Gao, Z.P.; Yan, H.X.; Ning, H.P.; Reece, M.J. Ferroelectricity of $Pr_2Ti_2O_7$ ceramics with super high Curie point. *Adv. Appl. Ceram.* **2013**, *112*, 69–74. [CrossRef]

12. Yan, H.; Ning, H.; Kan, Y.; Wang, P.; Reece, M.J. Piezoelectric Ceramics with Super-High Curie Points. *J. Am. Ceram. Soc.* **2009**, *92*, 2270–2275. [CrossRef]

13. Ning, H.; Yan, H.; Reece, M.J. Piezoelectric Strontium Niobate and Calcium Niobate Ceramics with Super-High Curie Points. *J. Am. Ceram. Soc.* **2010**, *93*, 1409–1413. [CrossRef]

14. Nanamatsu, S.; Kimura, M. Ferroelectric Properties of $Ca_2Nb_2O_7$ Single Crystal. *J. Phys. Soc. Jpn.* **1974**, *36*, 1495. [CrossRef]

15. Nanamats, S.; Kimura, M.; Doi, K.; Matsushi, S.; Yamada, N. A New ferroelectric: $La_2Ti_2O_7$. *Ferroelectrics* **1974**, *8*, 511–513. [CrossRef]

16. Kimura, M.; Nanamatsu, S.; Kawamura, T.; Matsushita, S. Ferroelectric, Electrooptic and Piezoelectric Properties of $Nd_2Ti_2O_7$ Single Crystal. *Jpn. J. Appl. Phys.* **1974**, *13*, 1473–1474. [CrossRef]

17. Gao, Z.; Lu, C.; Wang, Y.; Yang, S.; Yu, Y.; He, H. Super Stable Ferroelectrics with High Curie Point. *Sci. Rep.* **2016**, *6*, 24139. [CrossRef]

18. Barrow, D.; Petroff, T.; Sayer, M. Thick ceramic coatings using a sol gel based ceramic-ceramic 0–3 composite. *Surf. Coat. Technol.* **1995**, *76*, 113–118. [CrossRef]

19. Barrow, D.A.; Petroff, T.E.; Tandon, R.P.; Sayer, M. Characterization of thick lead zirconate titanate films fabricated using a new sol gel based process. *J. Appl. Phys.* **1997**, *81*, 876–881. [CrossRef]

20. Kobayashi, M.; Olding, T.R.; Zou, L.; Sayer, M.; Jen, C.K.; Rehmen, A.U. Piezoelectric Thick Film Ultrasonic Transducers Fabricated by a Spray Technique. In Proceedings of the IEEE Ultrasonics Symposium, San Juan, PR, USA, 22–25 October 2000; pp. 985–989.

21. Kobayashi, M.; Jen, C.K. Piezoelectric thick bismuth titanate/lead zirconate titanate composite film transducers for smart NDE of metals. *Smart Mater. Struct.* **2004**, *13*, 951–956. [CrossRef]

22. Kobayashi, M.; Jen, C.K.; Nagata, H.; Hiruma, Y.; Tokutsu, T.; Takenaka, T. 10F-5 Integrated Ultrasonic Transducers above 500°C. In Proceedings of the 2007 IEEE Ultrasonics Symposium, New York, NY, USA, 28–31 October 2007; pp. 953–956.

23. Kobayashi, M.; Jen, C.K.; Ono, Y.; Krüger, S. Lead-Free Thick Piezoelectric Films as Miniature High Temeprature Ultrasonic Transducer. In Proceedings of the IEEE Ultrasonic Symposium, Montreal, QC, Canada, 23–27 August 2004; pp. 910–913.

24. Kobayashi, M.; Ono, Y.; Jen, C.K.; Cheng, C.C. High-Temperature Piezoelectric Film Ultrasonic Transducers by a Sol-Gel Spray Technique and Their Application to Process Monitoring of Polyymer Injection Molding. *IEEE Sens. J.* **2006**, *6*, 55–62. [CrossRef]

25. Searfass, C.T.; Baba, A.; Tittmann, B.R.; Agrawal, D.K.; Thompson, D.O.; Chimenti, D.E. Fabrication and testing of microwave sintered sol-gel spray-on bismuth titanate-lithium niobate based piezoelectric composite for use as a high temperature (>500 °C) ultrasonic transducer. In *Review of Progress in Quantitative Nondestructive Evaluation*; Springer Science & Business Media: Berlin/Heidelberg, Germany, 2010; pp. 1035–1042.

26. Searfass, C.T.; Tittmann, B.R.; Agrawal, D.K. Sol-gel Deposited Thick Film Bismuth Titanate Based Transducer Achieves Operation of 600 C. In *Review of Progress in Quantitative Nondestructive Evaluation*; Springer Science & Business Media: Berlin/Heidelberg, Germany, 2010; pp. 1751–1758.

27. Searfass, C.T.; Parks, D.A.; Baba, A.; Tittmann, B.R. Testing of Single Crystal and Oriented Polycrystalline Piezoelectric Materials for Use as High Temperature (>500 degrees C) Ultrasonic Transducers. Presented at the 12th International Symposium on Nondestructive Characterization of Materials (NDCM-XII), Blacksburg, VA, USA, 19–24 June 2011.

28. Searfass, C.T.; Baba, A.; Tittmann, B.R.; Agrawal, D.K. Spray-on Piezoelectric Bismuth Titinate and Bismuth Titinate-Lithium Niobate Transducers for High Temperature (>500 degrees C) Ultrasonic Nondestructive Evaluation. Presented at the 12th International Symposium on Nondestructive Characterization of Materials (NDCM-XII), Blacksburg, VA, USA, 19–24 June 2011.

29. Cyphers, R.L. Low Frequency Testing of Ultrasonic Tranducers fabricated with a sol-gel Spray-on Procedure. Master's Thesis, The Pennsylvania State University, Stecker Ridge, PA, USA, 2012.

30. Sinclair, A.N.; Chertov, A.M. Radiation endurance of piezoelectric ultrasonic transducers—A review. *Ultrasonics* **2014**, *57*, 1–10. [CrossRef]

31. Tittmann, B.R.; Reinhardt, B.; Parks, D.R. Nuclear Radiation Tolerance of Single Crystal Aluminum Nitride Ultrasonic Transducer. In Proceedings of the 2014 IEEE International Ultrasonics Symposium, Chicago, IL, USA, 3–6 September 2014. [CrossRef]

32. Parks, D.A.; Tittmann, B.R. Radiation tolerance of piezolectric bulk single crystal aluminum nitride. *IEEE Trans. Ultrason. Ferroelectr. Freq. Control* **2014**, *61*, 1216–1222. [CrossRef]

33. Reinhardt, B.; Daw, J.; Tittmann, B.R. Irradiation Testing of Piezoelectric (Aluminum Nitride, Zinc Oxide, Bismuth Titanate) and Magnetostrictive Sensors (Remendur and Galfenol). *IEEE Trans. Nucl. Sci.* **2018**, *65*, 533–538. [CrossRef]

34. Anon. Evaluation of Irradiation Test Results for Candiate NERVA. In *Piezoelectric Accelerometers, Ground Test Reactor Irradiation Test No. 22*; Aerojet Nuclear Systems Company: Huntsville, AL, USA, 1971.

35. Berger, L. *Semiconductor Materials*; CRC Press: Boca Raton, FL, USA, 1997.

36. Baranov, V.M.; Martynenko, S.P.; Sharapa, A.I. Durability of ZTL piezoceramic under the action of reactor radiation. *At. Energy* **1982**, *53*, 803–804. [CrossRef]

37. Brinkman, J.A. On the Nature of Radiation Damage in Metals. *J. Appl. Phys.* **1954**, *25*, 961. [CrossRef]

38. Broomfield, G. The effect of low-fluence neutron irradiation on silver-electroded lead-zirconate-titanate piezoelectric ceramics. *J. Nucl. Mater.* **1980**, *91*, 23–34. [CrossRef]

39. Dienes, G.J.; Vineyard, G.H. *Radiation Effects in Solids*; Interscience Publishers Inc.: New York, NY, USA, 1957.

40. Friedland, E. Influence of electronic stopping on amorphization energies. *Surf. Coat. Technol.* **2007**, *201*, 8220–8224. [CrossRef]

41. Glower, D.D.; Hester, D.L. Hysteresis Studies of Reactor-Irradiated Single-Crystal Barium Titanate. *J. Appl. Phys.* **1965**, *36*, 2175. [CrossRef]

42. Glower, D.D.; Hester, D.L.; Warnke, D.F. Effects of Radiation-Induced Damage Centers in Lead Zirconate Titanate Ceramics. *J. Am. Ceram. Soc.* **1965**, *48*, 417–421. [CrossRef]

43. Hobbs, L.W.; Clinard, F.W., Jr.; Zinkle, S.J.; Ewing, R.C. Radiation effects in ceramics. *J. Nucl. Mater.* **1994**, *216*, 291–321. [CrossRef]

44. Hobbs, L.W.; Jesurum, C.E.; Berger, B. Rigid Constraints in Amorphization of Singly and Multiply-Polytopic Structures. In *Rigidity Theory and Applications*; Thorpe, M.F., Duxbury, P.M., Eds.; Kluwer Academic/Plenium Publishers: Heidelberg, Germany, 2000; pp. 191–216.

45. Ito, Y.; Yasuda, K.; Ishigami, R.; Sasase, M.; Hatori, S.; Ohashi, K.; Tanaka, S.; Yamamoto, A. Radiation damage of materials due to high-energy ion irradiation. *Nucl. Instrum. Methods Phys. Res. Sect. B Beam Interact. Mater. Atoms* **2002**, *191*, 530–535. [CrossRef]

46. Kazys, R.; Voleisis, A.; Sliteris, R.; Mazeika, L.; Van Nieuwenhove, R.; Kupschus, P.; Abderrahim, H.A. High temperature ultrasonic transducers for imaging and measurements in a liquid Pb/Bi eutectic alloy. *IEEE Trans. Ultrason. Ferroelectr. Freq. Control* **2005**, *52*, 525–537. [CrossRef] [PubMed]

47. Lefkowitz, I. Radiation-Induced Changes in the Ferroelectric Properties of Some Barium Titanate-Type Materials. *J. Phys. Chem. Solids* **1958**, *10*, 169–173. [CrossRef]

48. Meldrum, A.; Boatner, L.; Weber, W.; Ewing, R.; Boatner, L.; Weber, W. Amorphization and recrystallization of the ABO3 oxides. *J. Nucl. Mater.* **2002**, *300*, 242–254. [CrossRef]

49. Meleshko, Y.P.; Babaev, S.V.; Karpechko, S.G.; Nalivaev, V.I.; Safin, Y.A.; Smirnov, V.M. Studying the electrophysical parameters of piezoceramics of various types in an IVV-2M reactor. *At. Energy* **1984**, *57*, 544–548. [CrossRef]

50. Miclea, C.; Miclea, C.F.; Spanulescu, I.; Cioangher, M.; Tănăsoiu, C. Effect of neutron irradiation on some piezoelectric properties of PZT type ceramics. *J. Phys.* **2005**, *128*, 115–120. [CrossRef]

51. Primak, W.; Anderson, T.T. Metamimictization of Lithium Niobate by Thermal Neutrons. *Nucl. Technol.* **1975**, *23*, 235–248.

52. Rempe, J.; MacLean, H.; Schley, R.; Hurley, D.; Daw, J.; Taylor, S.; Smith, J.; Svoboda, J.; Kotter, D.; Knudson, D.; et al. *Strategy for Developing New in-Pile Instrumentation to Support Fuel Cycle Research and Development*; Idaho National Laboratory: Idaho Falls, ID, USA, 2011.

53. Sickafus, K.E.; Kotomin, E.A.; Ub, B.P. *Radiation Effects in Solids (Proceedings of the NATO Advanced Study Institute on Radiation Effects in Solids)*; NATO Science Series; Springer: Erice, Italy, 2007.

54. Szenes, G. Ion-induced amorphization in ceramic materials. *J. Nucl. Mater.* **2005**, *336*, 81–89. [CrossRef]

55. Szenes, G. Thermal spike analysis of ion-induced tracks in semiconductors. *Nucl. Instrum. Methods Phys. Res. Sect. B Beam Interact. Mater. Atoms* **2011**, *269*, 2075–2079. [CrossRef]

56. Thomas, R.L. *Vibration Instrumentation for Nuclear Reactors*; Endevco Technical Report 258; Los Alamos National Lab.: Los Alamos, NM, USA, 1973; pp. 1–8.

57. Trachenko, K. Understanding resistance to amorphization by radiation damage. *J. Phys. Condens. Matter* **2004**, *16*, R1491–R1515. [CrossRef]

58. Trachenko, K.; Pruneda, J.M.; Artacho, E.; Dove, M.T.; Pruneda, M. How the nature of the chemical bond governs resistance to amorphization by radiation damage. *Phys. Rev. B* **2005**, *71*, 1–5. [CrossRef]

59. Yano, T.; Inokuchi, K.; Shikama, M.; Ukai, J.; Onose, S.; Maruyama, T. Neutron irradiation effects on isotope tailored aluminum nitride ceramics by a fast reactor up to 2×10^{26} n/m^2. *J. Nucl. Mater.* **2004**, *329*, 1471–1475. [CrossRef]

60. Zhang, S.; Yu, F. Piezoelectric Materials for High Temperature Sensors. *J. Am. Ceram. Soc.* **2011**, *94*, 3153–3170. [CrossRef]

61. Kong, X.Y.; Ding, Y.; Yang, R.; Wang, Z.L. Single-Crystal Nanorings Formed by Epitaxial Self-Coiling of Polar Nanobelts. *Science* **2004**, *303*, 1348–1351. [CrossRef]

62. Parks, D.A.; Reinhardt, B.T.; Tittmann, B.R. Piezoelectric Material for Use in a Nuclear Reactor Core. Presented at the Review of Progress in Quantitative Nondestructive Evaluation, Burlington, VT, USA, 17–22 July 2011; pp. 1633–1639.

63. Reinhardt, B.T.; Parks, D.A.; Tittmann, B.R. Measurement of Irradiation Effects in Precipitate Hardened Aluminum Using Nonlinear Ultrasonic Principles (In-Situ). Presented at the Review of Progress in Quantitative Nondestructive Evaluation, Burlington, VT, USA, 17–22 July 2011; pp. 1648–1654.

64. Reinhardt, B.T.; Searfass, C.T.; Pheil, C.; Sinding, K.; Cyphers, R.; Tittmann, B.R. Fabrication of Bismuth Titanate-PZT Ceramic Transducers for High Temperature Applications. In Proceedings of the Review of Progress in Quantitative NDE, Denver, CO, USA, 15–20 July 2012.

65. Reinhardt, B.T.; Parks, D.A.; Tittmann, B.R. Selecting a Radiation Tolerant Piezoelectric Material for Nuclear Reactor Applications. In Proceedings of the Review of Progress in Quantitative NDE, Denver, CO, USA, 15–20 July 2012.

66. Daw, J.; Tittmann, B.; Reinhardt, B.; Kohse, G.; Ramuhalli, P.; Montgomery, R.; Chien, H.T.; Villard, J.F.; Palmer, J.; Rempe, J. Irradiation testing of ultrasonic transducers. In Proceedings of the 2013 3rd International Conference on Advancements in Nuclear Instrumentation, Measurement Methods and Their Applications (ANIMMA), Marseille, France, 23–27 June 2013.

67. Sinding, K.; Searfass, C.; Malarich, N.; Reinhardt, B.; Tittmann, B.R. High temperature ultrasonic transducers for the generation of guided waves for non-destructive evaluation of pipes. In Proceedings of the 40th Annual Review of Progress in Quantitative Nondestructive Evaluation: Incorporating the 10th International Conference on Barkhausen Noise and Micromagnetic Testing, Baltimore, MD, USA, 21–26 July 2013; pp. 302–307.

68. Daw, J.E.; Rempe, J.L.; Palmer, J.; Tittmann, B.R.; Reinhardt, B.; Kohse, G.; Ramuhalli, P.; Chien, H.T. Assessment of Survival or Ultrasonic Transducers under Neutron Irradiation. In *Nuclear Fuels and Materials Spotlight*; Byrd, L., Ed.; Idaho National Laboratory: Idaho Falls, ID, USA, 2014; Volume 4.

69. Ledford, K.R.; Sinding, K.; Tittmann, B. Active and passive monitoring of valve bodies utilizing spray-on transducer technology. *J. Acoust. Soc. Am.* **2015**, *137*, 2234. [CrossRef]

70. Reinhardt, B.; Tittmann, B.; Rempe, J.; Daw, J.; Kohse, G.; Carpenter, D.; Ames, M.; Ostrovsky, Y.; Ramuhalli, P.; Montgomery, R.; et al. Progress towards developing neutron tolerant magnetostrictive and piezoelectric transducers. In *41st Annual Review of Progress in Quantitative Nondestructive Evaluation*; Chimenti, D.E., Bond, L.J., Eds.; AIP Publishing: Melville, NY, USA, 2015; Volume 1650, pp. 1512–1520.

71. Reinhardt, B.; Tittmann, B.R.; Suprock, A. Nuclear Radiation Tolerance of Single Crystal Aluminum Nitride Ultrasonic Transonic. In Proceedings of the 2015 ICU International Congress on Ultrasonics, Metz, France, 15–24 May 2015; Volume 70, pp. 609–613.

72. Tittmann, B.R. Transducers for In-Pile Ultrasonic Measurement of the Evolution of Fuels and Materials. In *Nuclear Science User Facilities 2014 Annual Report*; Department of Energy: Idaho Falls, ID, USA, 2015.

73. Reinhardt, B.T.; Suprock, A.; Tittmann, B. Testing piezoelectric sensors in a nuclear reactor environment. In *43rd Annual Review of Progress in Quantitative Nondestructive Evaluation*; Chimenti, D.E., Bond, L.J., Eds.; AIP Publishing: Melville, NY, USA, 2017.

74. Giot, M.; Vermeeren, L.; Lyoussi, A.; Reynard-Carette, C.; Lhuillier, C.; Mégret, P.; Deconinck, F.; Gonçalves, B.S. Nuclear instrumentation and measurement: A review based on the ANIMMA conferences. *EPJ Nucl. Sci. Technol.* **2017**, *3*, 33. [CrossRef]

75. Lissenden, C.J.; Tittmann, B.R. Temperature Resistant Spray-on Piezoelectric Transducers for Materials Characterization with Ultrasonic-Guided Waves. In *43rd Annual Review of Progress in Quantitative Nondestructive Evaluation*; Chimenti, D.E., Bond, L.J., Eds.; AIP Publishing: Melville, NY, USA, 2017; Volume 1806, p. 05005.

76. Searfass, C.T.; Pheil, C.; Sinding, K.; Tittmann, B.R.; Baba, A.; Agrawal, D.K. Bismuth Titanate Fabricated by Spray-on Deposition and Microwave Sintering for High-Temperature Ultrasonic Transducers. *IEEE Trans. Ultrason. Ferroelectr. Freq. Control* **2016**, *63*, 139–146. [CrossRef]

77. Sinding, K.M.; Orr, A.; Breon, L.; Tittmann, B.R. Effect of sintering temperature on adhesion of spray-on piezoelectric transducers. *J. Sens. Sens. Syst.* **2016**, *5*, 113–123. [CrossRef]

78. Suprock, A.D.; Tittmann, B.R. Development of high temperature capable piezoelectric sensors. In *43rd Annual Review of Progress in Quantitative Nondestructive Evaluation*; Chimenti, D.E., Bond, L.J., Eds.; AIP Publishing: Melville, NY, USA, 2017; Volume 1806, p. 05007.

79. Malarich, N.; Lissenden, C.J.; Tittmann, B.R. Field Deployable Processing Methods for Stay-in-Place Ultrasonic Transducers. In *43rd Annual Review of Progress in Quantitative Nondestructive Evaluation*; Chimenti, D.E., Bond, L.J., Eds.; AIP Publishing: Melville, NY, USA, 2017; Volume 1806, p. 05005.

80. Xu, J.L.; Batista, C.F.G.; Tittmann, B.R. Practical ultrasonic transducers for high-temperature applications using bismuth titanate and Ceramabind 830. In Proceedings of the 44th Annual Review of Progress in Quantitative Nondestructive Evaluation, Provo, UT, USA, 16–21 July 2017.

81. Trivedi, Y.; Tittmann, B.R.; Batista, C.; Xu, J.; Lissenden, C. Field deployable spray-on ultrasonic coatings for high-temperature applications. *AIP Conf. Proc.* **2019**, *2102*, 060008. [CrossRef]

82. Cho, H. Toward Robust SHM and NDE of Plate-like Structures using Nonlinear Guided Wave Features. Ph.D. Thesis, The Pennsylvania State University, State College, PA, USA, 2017.

83. Gatsa, O.; Combette, P.; Rosenkrantz, E.; Fourmentel, D.; Destouches, C.; Ferrandis, J.Y. High-Temperature Ultrasonic Sensor for Fission Gas Characterization in MTR Harsh Environment. *IEEE Trans. Nucl. Sci.* **2018**, *65*, 2448–2455.

84. Very, F.; Rosenkrantz, E.; Fourmentel, D.; Destouches, C.; Villard, J.F.; Combette, P.; Ferrandis, J.Y. Acoustic sensors for fission gas characterization in MTR harsh environment. *Physics Procedia* **2015**, *70*, 292–295.

Technical Note

Pb(Mg$_{1/3}$Nb$_{2/3}$)-PbTiO$_3$-Based Ultrasonic Transducer for Detecting Infiltrated Water in Pressurized Water Reactor Fuel Rods

Geonwoo Kim [1,2], Namkyoung Choi [1,3], Yong-Il Kim [1,4] and Ki-Bok Kim [1,3,*]

[1] Department of Science of Measurement, University of Science and Technology, 217, Gajeong-ro, Yuseong-gu, Daejeon 34113, Korea; Geonwoo.Kim@usda.gov (G.K.); gnokd@kriss.re.kr (N.C.); yikim@kriss.re.kr (Y.-I.K.)

[2] Environmental Microbial and Food Safety Laboratory, Agricultural Research Service, U.S. Department of Agriculture, Powder Mill Rd. Bldg. 303, BARC-East, Beltsville, MD 20705, USA

[3] Center for Safety Measurement, Korea Research Institute of Standards and Science, 267, Gajeong-ro, Yuseong-gu, Daejeon 34113, Korea

[4] Center for Convergence Property Measurement, Korea Research Institute of Standards and Science, 267, Gajeong-ro, Yuseong-gu, Daejeon 34113, Korea

* Correspondence: kimkibok@kriss.re.kr; Tel.: +82-42-868-5650

Received: 18 April 2019; Accepted: 11 June 2019; Published: 13 June 2019

Abstract: In this study, a high-sensitivity Pb(Mg$_{1/3}$Nb$_{2/3}$)O$_3$−PbTiO$_3$ (PMN-PT)-based ultrasonic transducer was developed for detecting defective pressurized water reactor (PWR) fuel rods. To apply the PMN-PT substance to nuclear power plant facilities, given the need to guarantee their robustness against radioactive materials, the effects of neutron irradiation on PMN-PT were investigated. As a result, the major piezo-electric constants of PMN-PT, such as the electrical impedance, dielectric constant, and piezo-electric charge constant, were found to vary within acceptable ranges. This means that the PMN-PT could be used as the piezo-electric material in the ultrasonic transducer for nuclear power plants. The newly developed ultrasonic transducer was simulated using a modified KLM model for the through-transmission method and fabricated under the same conditions as in the simulation. The through-transmitted waveforms of normal and defective PWR fuel rods were obtained and compared with simulated results in the time and frequency domains. The response waveforms of the newly developed ultrasonic transducer for pressurized water reactor (PWR) fuel rods showed good agreement with the simulation outcome and could clearly detect defective specimens with high sensitivity.

Keywords: nuclear power plants; pressurized water reactor fuel rods; ultrasonic transducer; PMN-PT; neutron irradiation; non-destructive evaluation

---

1. Introduction

A pressurized water reactor (PWR) fuel rod typically consists of sintered uranium dioxide pellets stacked into thin zircaloy tubes (~ 4–5 m), which are sealed by welding at both ends. The zircaloy tubes are the first protection against leakage of the dangerously radioactive fission products. If deterioration of a PWR fuel rod occurs in coolant, its fission products are leaked and the defective PWR fuel rod is filled with cooling water. Therefore, detecting the presence of infiltrated water in the PWR fuel rods, which indicates cladding failure and lack of mechanical integrity, is very important for the safety of PWRs. Ultrasonic transducers based on lead zirconate titanate (PZT) ceramics have been widely used in the non-destructive evaluation (NDE) of nuclear power plant (NPP) components as well as in various other industrial fields due to their high performance and ease of manufacture [1–4].

However, the PZT-based ultrasonic transducers have some limitations for application to the high-attenuation materials used in NPPs because of its low piezo-electric constant [5,6]. In contrast, Pb$(Mg_{1/3}Nb_{2/3})O_3-PbTiO_3$ (PMN-PT) single crystal is currently considered one of the most important among the emerging piezo-electric materials for high-sensitivity ultrasonic transducers. This is because of its superior piezo-electric constants, including exceptional piezo-electric charge constant ($d_{33} \geq 1500$ pC/N), high electro-mechanical coupling factor ($k_{33} \sim 0.9$), and superior dielectric constant ($\varepsilon_r \sim 5000$) [5–10]. In addition, these piezo-electric constants can provide good sensitivity and bandwidth for ultrasonic imaging techniques [10–12]. In spite of these advantages, PMN-PT is not widely used for NDE of NPP applications because there has been little conformance verification on its radiation and high-temperature resistance. Much research about the effects of various kinds of irradiation on ferro-, pyro-, and piezo-electric elements have been reported [13–15]. However, to the best of our knowledge, only a few studies on the effects of radiation on PMN-PT have been conducted [7,16]. In our previous study, we investigated the effects of neutron irradiation on PMN-28%PT for NPP applications [16]. Increasing the neutron radiation dose to 160 Nm/g (Mrad) changes the major piezo-electric constants, which can affect the overall performance of ultrasonic transducers. For this reason, a number of qualities such as the dielectric constant ε_r, piezo-electric coefficient d_{33}, electrical impedance |Z|, and electrical resistance (ohm, Ω) were measured and analyzed. In addition, X-ray diffraction (XRD) was used to analyze the changes of the piezo-electric constant values. Angadi et al. reported the radiation response of PMN-PT to high-energy heavy ions (50 MeV Li^{3+}, fluence 1×10^{13}–1×10^{14} ions/cm^2), in terms of its structural, dielectric, and piezo-electric constants at high temperature (180 °C) [7]. The current NDT methods to evaluate the extent of failure of in-service or post irradiated fuel rods are eddy current testing (ECT) and ultrasonic testing (UT) [17]. UT testing for in-service NPPs is typically conducted by its maintenance periodic testing activities after reactor shutting down and the temperature of coolant system is cooled down below 93 °C [1,18,19]. The two experimental studies clearly demonstrated that the piezo-electric constant and crystallinity of the material showed only slight changes (within acceptable ranges) under harsh environments and that variation of the major piezo-electric constants hardly affected the overall piezo-electric performance of PMN-PT.

Therefore, we developed a PMN-PT-based high-sensitivity 5 MHz ultrasonic transducer for detecting defective PWR fuel rods of in-service PWRs. To accomplish this, we designed the ultrasonic transducer and simulated its response waveforms. For the ultrasonic transducer design and simulation, various models based on equivalent circuits have been developed such as the Mason, Redwood, and KLM models [20–22]. Among these models, the KLM model was selected for this study due to its simple acoustic transmission nodes and transfer functions that operate in a more physically intuitive manner than in the others [23–26]. The KLM model can simulate the thickness-vibration mode of a piezo-electric material, from which its ultrasonic pulse-echo response signals can be obtained. However, because the technique for detection of the permeated water is the through-transmission method, in this study, we used a modified KLM model developed in a previous study for the ultrasonic through-transmission method [25]. Then, the through-transmitted waveforms of defective PWR fuel rods were simulated. To verify the outcome, a prototype ultrasonic transducer was fabricated, and the experimental set up duplicated the same conditions used in the simulation. Then, the through-transmitted waveforms of normal and defective PWR fuel rods were obtained in time and frequency domains and compared with the simulation results.

2. Neutron Irradiation Effects on PMN-28%PT

In our previous study, according to neutron irradiation dose on PMN-28%PT (Ibulephotonics Co., Ltd., Republic of Korea), changes of its electrical impedance, |Z|, dielectric constant, ε_r, and piezoelectric charge constant, d_{33} were measured by an impedance and gain-phase analyzer (HP 4194A, Hewllet Packard Co., Ltd., Palo Alto, CA, USA), LCR meter (4263B LCR meter, Agilent Co., Ltd, Santa Clara, CA, USA), and d_{33}-meter (ZJ-6B PIEZO d_{33}/d_{31} meter, Institute of Acoustics Chinese Academy of Sciences, Beijing, China), respectively, and its results were analyzed by X-ray diffractometry (D/Max 2200, Rigaku Co., Ltd., Tokyo, Japan). Also, obtained XRD pattern data were sorted by search–match method [16]. The number of PMN-28%PT samples was three and those were irradiated by Cf-252 neutron source from 0 to 160 Nm/g (Mrad) at a neutron dose rate of 50 Nm/g/h (Mrad/h) at room temperature. The center frequency and diameter of PMN-28%PT samples were 1 MHz and 1.8 mm. Figure 1 shows the results of the neutron irradiation effects on three PMN-28%PT samples.

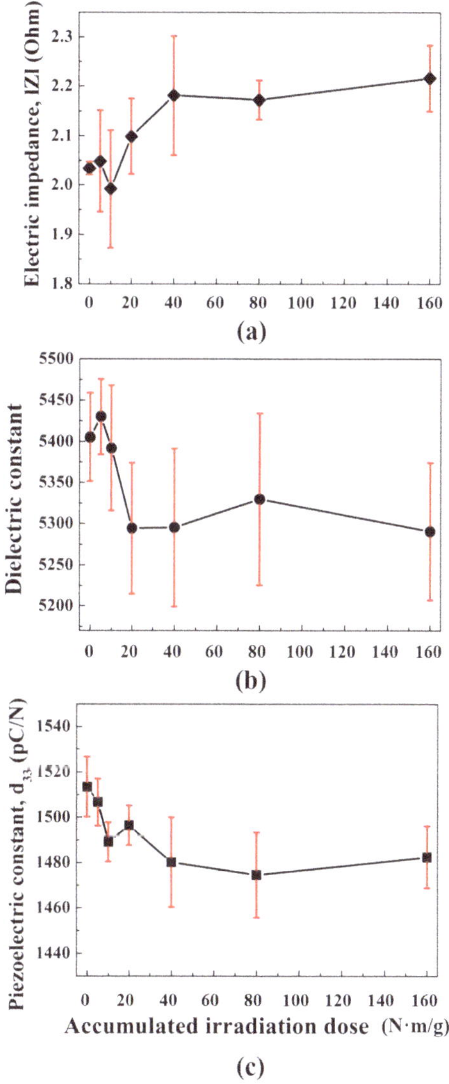

Figure 1. Results of the neutron irradiation effects on three PMN-28%PT samples: (**a**) Electrical impedance, |Z|; (**b**) dielectric constant, ε_r; (**c**) piezo-electric charge constant, d_{33}.

In Figure 1, the black lines are the average of three samples and red vertical lines represented the 95% confidence intervals. In Figure 1a, the electrical impedance increased up to 40 Nm/g dose and after 40 Nm/g, it maintained constant values. In Figure 1b, permittivity slightly increased and decreased rapidly up to 20 Nm/g. Then, it represented uniform values. In Figure 1c, d_{33} decreased up to 40 Nm/g and maintained uniform values up to 160 Nm/g. To analyze the causes of those property changes, its XRD pattern analysis was performed. The measurement conditions of the XRD experiment is summarized in Table 1, and Figure 2 shows obtained XRD patterns of three PMN-28%PT samples.

Table 1. Measurement conditions of X-ray diffractometry.

Parameters	Units	Values
X-ray source	.	Copper (Cu)
Load	kV	40
Sample mode	.	reflection
Scan mode	.	step scan
Scan range (2θ)	Degree, °	3600
Scan interval (2θ)	Degree, °	0.02
Monochromator	.	Curved graph
Detector	.	Scintillation Counter

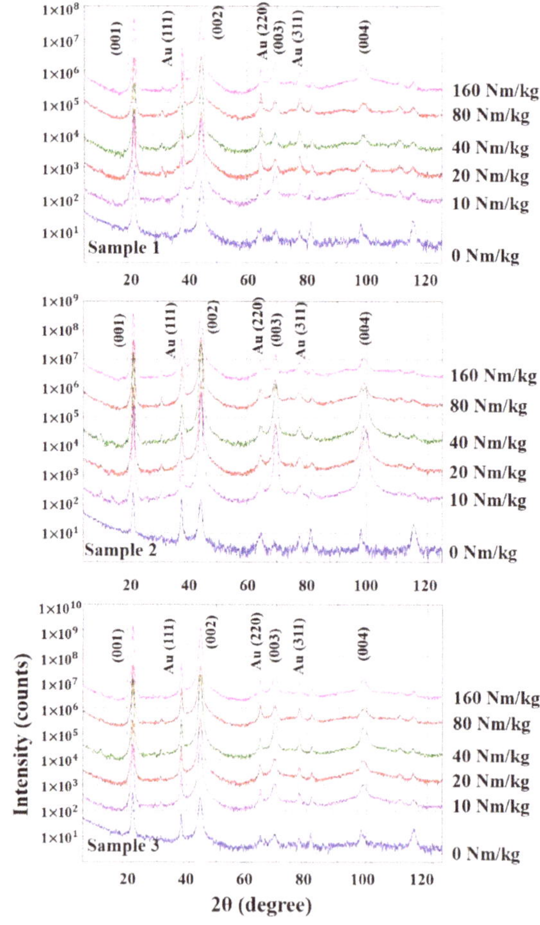

Figure 2. The intensity variation of three PMN-28%PT XRD patterns at (001) plane according to irradiation does levels.

In Figure 2, XRD pattern of (001) plane of PMN-28%PT and Au (111) peak appeared very high. (111) peak is the electrode (Au) of PMN-28%PT. From the obtained XRD data for analyzing the formation of the PMN-28%PT samples in single phase, the radioactive irradiation affected the material structure of PMN-28%PT. The crystal structures of PMN-28%PT, which can be described by [(1-x)Pb(Mg$_{1/3}$Nb$_{2/3}$)O$_3$-xPbTiO$_3$] are rhombohedral, monoclinic, tetragonal, and cubic phase at room temperature in accordance with the amount of PbTiO$_3$ content [16]. From Figure 2, the crystal structure of PMN-28%PT samples showed tetragonal phase. The polarization of PMN-28%PT, which has tetragonal phase, can be influenced by the degree of off-center of atoms occupied the (00l) reflection. In addition, gradual increase of X-ray diffracted intensities for (002), (003), and (004) reflections means the increase of reflection plane density due to atoms occupied the (00l) toward the center position. Consequently, the degradation of d_{33} and dielectric constant was affected by this movement of atoms. In the view point of qualitative analysis, its phase was not changed. However, the full width at half maximum (FWHM), which represents its crystallinity, irregularly varied according to the irradiation does as shown in Figure 3.

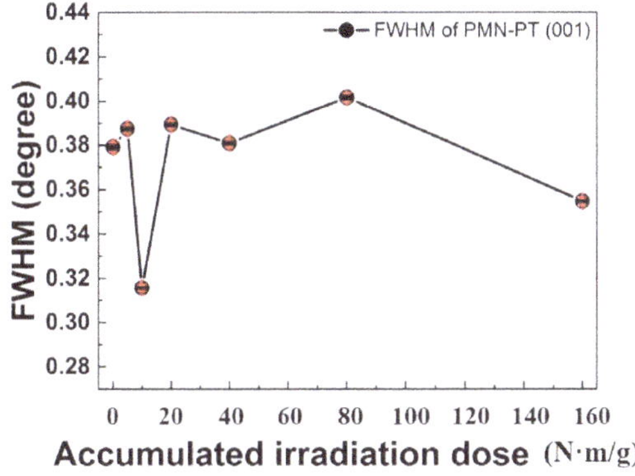

Figure 3. Variation of full width at half maximum (FWHM) at (001) plane of PMN-28%PT in accordance with the accumulated neutron irradiation does.

In Figure 3, the variation of the FWHM diffraction peak reflects the change of its crystallinity with an increase in the radiation dose. Also, the irregularity of its crystallinity is one of the important major causes that develop defect density such as sample dislocation. This phenomenon can cause the permanent dipole moment decreases in the PMN-28%PT crystal structure. From the above analysis, we concluded that the variation of major piezo-electric constants can be caused by neutron irradiation dose [16]. Although there was slight decrease in the PMN-28%PT crystallinity according to increasing neutron irradiation dose, the major piezo-electric constants, which can decide the ultrasonic transducer performance, such as electrical impedance, dielectric constant, and piezoelectric charge constant, varied within acceptable ranges as shown in Figure 1. Therefore, the PMN-28%PT single crystal could be a promising piezo-electric material for NPPs application.

3. Ultrasonic Transducer Design

3.1. Structure of Ultrasonic System and Transducer

The ultrasonic testing system for detecting defective PWR fuel rods consisted of transmitting (Tx) and receiving (Rx) ultrasonic transducers, a test specimen, an oscilloscope, a pulse generator, and a receiver. Unlike a pulse-echo method, which uses one ultrasonic transducer, the through-transmission method utilizes both Tx and Rx ultrasonic transducers. The conceptual diagram of a device inspecting PWR fuel rods is shown in Figure 4.

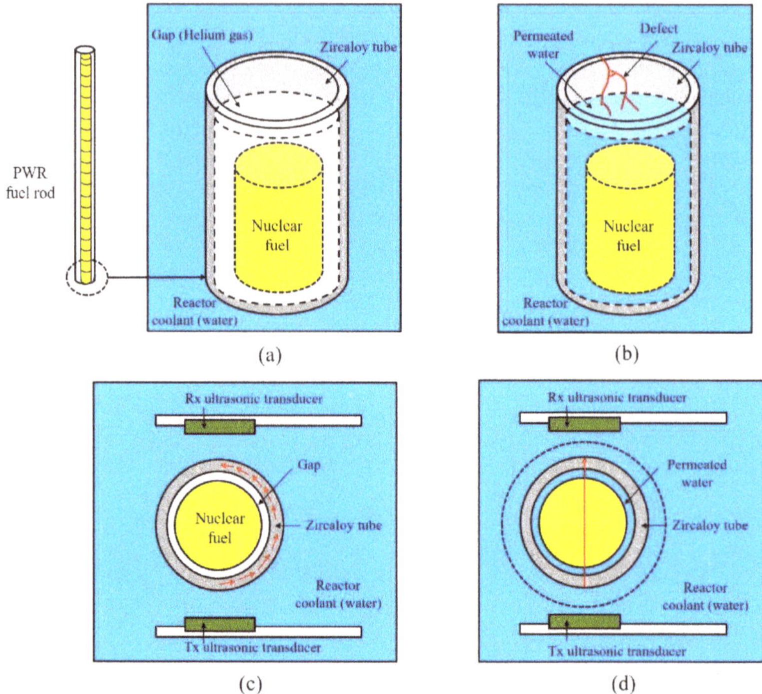

Figure 4. Conceptual diagram of sound and defective pressurized water reactor (PWR) fuel rods and its ultrasonic inspection: (**a**) Sound PWR fuel rod; (**b**) defective PWR fuel rod with permeated water from reactor coolant; (**c**) ultrasonic inspection of sound PWR fuel rod and its ultrasonic path, which is trapped in PWR fuel rod; (**d**) ultrasonic inspection of defective PWR fuel rod and through-transmitted ultrasonic path.

Figure 4a,b shows the sound and defective PWR fuel rods. Uranium powder baked into cylindrical yellow pellets were stacked into a thin zircaloy tube. The outer diameter and thickness of PWR fuel rod is 10 mm and 1 mm, respectively. There is a gap between the tube and the pellets, and the gap is filled with helium gas to improve heat transport from the nuclear fuel to the reactor coolant [2–4]. However, in the case of defective fuel rod, the cooling water can be permeated through the zircaloy tube and the gap is filled with the water instead of the gas. When PWR fuel rods are inspected, the main way to receive the elastic waves generated by the Tx ultrasonic transducer is via the tube, as described in Figure 4c, and the distance between both ultrasonic transducers is about 12 mm. If defects occur, cooling water fills the gas gap and the ultrasonic waves are directly transmitted to the Rx ultrasonic transducer, as represented in part Figure 4d. Therefore, the NDE of the PWR fuel rods is conducted by analyzing the through-transmitted ultrasonic signals propagated within the zircaloy tubes.

The structure of the ultrasonic transducer includes the piezo-electric element, front matching layers, metal cases, and backing materials. The piezo-electric material transforms the acoustic energy into electrical signals, and vice versa. The backing materials act like a mechanical damper that absorbs the ultrasonic waves generated on the back side of the piezo-electric element. It also determines the bandwidth of the response waveforms by controlling their acoustic impedance [25–28]. The front matching layer is located between the piezo-electric element and a test specimen and serves as an acoustic impedance matching filter to minimize loss of the acoustic energy and to control the phase of the transmitted ultrasonic signals [5,6,25–28]. The structure of the PMN-28%PT ultrasonic transducer developed for NDE of PWR fuel rods is shown in Figure 5.

Figure 5. Structure of the PMN-28%PT ultrasonic transducer developed for non-destructive evaluation (NDE) of PWR fuel rods.

To serve as the acoustic matching filter, the acoustic impedance of the front matching layer must be derived, and its ideal acoustic impedance is defined in Equation (1) [27,28].

$$Z_2 = \sqrt{Z_1 \cdot Z_3},\qquad(1)$$

where Z_1 and Z_3 are the acoustic impedance of the piezo-electric material and medium (a test specimen), respectively. Because the PWR fuel rods are in the coolant, the acoustic impedance of the medium was assumed to be that of water (1.5×10^6 kg/m^{2}s). The calculated acoustic impedance for the front matching layer was 6.6×10^6 kg/m^{2}s and poly methyl methacrylate (4×10^6 kg/m^{2}s) was selected because of its ease of mechanical working and low acoustic attenuation coefficient. In addition, the appropriate thickness of the front matching layer is one quarter of its wavelength at the operating frequency of the ultrasonic transducer, and the calculated thickness of the poly methyl methacrylate was about 0.1 mm [27,28]. In addition, to determine the acoustic impedance of the backing material, backing materials with various acoustic impedance were simulated using the KLM model. From this, 8×10^6 kg/m^{2}s was selected as the acoustic impedance of the ultrasonic transducer, with about 60% bandwidth as in our previous studies [5,6,25,26].

3.2. KLM Model Modified for the Through-Transmission Method

In our previous study, according to the neutron irradiation dose on PMN-28%PT, changes of its electrical impedance, $|Z|$, the dielectric constant, ε_r, and the piezo-electric charge constant, d_{33} were measured and analyzed because they are the crucial factors that affect the sensitivity and bandwidth of ultrasonic transducers [16,27–29]. The KLM modeling codes for ultrasonic through-transmission method are included in Supplementary file 1. To improve the electrical transmission loss and to control the bandwidth, the electrical impedance of piezo-electric materials is used to design the electric tuning circuit, which is located between the ultrasonic transducer and connected ultrasonic system. In particular, the dielectric constant and piezo-electric charge constant can directly affect the sensitivity of an ultrasonic transducer [20–22]. From XRD results, the major properties of PMN-28%PT are known to change its crystallinity with increasing neutron dose (from 0 to 16 Mrad). The variation was within an acceptable range (about ±5%) [16]. Therefore, we concluded that PMN-28%PT could be applied for NDE in NPPs. The material properties of the PMN-28%PT used in this study are summarized in Table 1 and the same properties were also used for the simulation.

In Table 2, the size of the PMN-28%PT was 3 × 3 mm because it should be smaller than the diameter of the PWR fuel rods (10 mm) to prevent the acoustic energy generated by the Tx ultrasonic transducer from being directly transmitted to the Rx ultrasonic transducer. The dielectric constant and piezo-electric constant were measured using an LCR meter (4263B LCR meter, Agilent Co., Ltd., Santa Clara, CA, USA) and d_{33}-meter (ZJ-6B PIEZO d_{33}/d_{31} meter (Institute of Acoustics Chinese Academy of Sciences, Beijing, China), respectively. The acoustic impedance of the PMN-28%PT, including its density and sound speed, was determined using XRD diffraction analysis and the ultrasonic pulse-echo method. The curie temperature of PMN-28%PT is 160 °C and the major piezo-electric constants such as dielectric constant and d_{33}, which decide the sensitivity of ultrasonic transducer showed acceptable ranges in other studies [7,10]. To simulate the through-transmitted waveforms, we developed the modified KLM model: A schematic including a test specimen and ultrasonic transducers (Tx and Rx) is shown in Figure 6 [25].

Table 2. Measured material properties of PMN-28%PT single crystal.

Parameters	Units	Values
Density	Kg/m^3	8000
Area of piezo-element	mm^2	9
Dielectric constant at 1 kHz	ε_r	5400
Piezo-electric coefficient, d_{33}	pC/N	1515
Longitudinal velocity	m/s	3600
Acoustic impedance	kg/m^{2}s	29×10^6
Operating frequency	MHz	5
Electro-mechanical coupling coefficient	k_{33}	0.9
Curie temperature	$^\circ$C	160

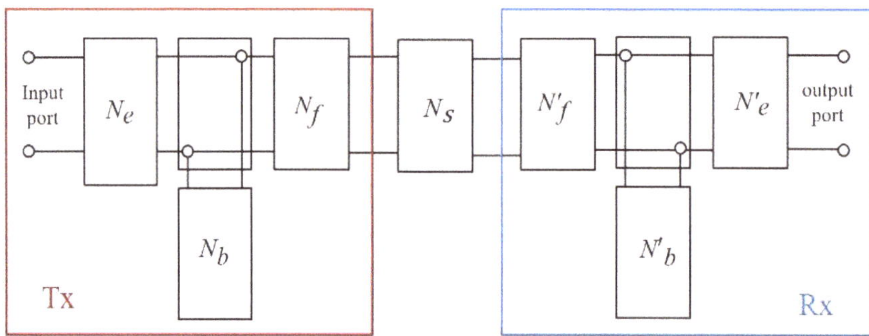

Figure 6. Schematic diagram of the modified KLM model for the through-transmission method.

In Figure 6, the components of the KLM model modified for the through-transmission method consist of equivalent circuits of the electro-mechanical coupling circuit N_e, backing material N_b, front matching layer N_f, and the medium Ns (a test specimen). The Tx ultrasonic transducer is excited by an excitation signal at the input port and the signal is converted to an ultrasonic wave described as frequency-domain matrices by an electro-mechanical coupling circuit. Then, the ultrasonic signal is tuned by the backing and the front matching layer and transmitted to the Rx ultrasonic transducer through the medium. Each component of the modified KLM model is described as a series of transfer matrices and the final through-transmitted outcome is obtained by a total transmission function in frequency domain. Finally, the result is transformed in time domain using inverse fast Fourier transform (IFFT) [22–25]. The excitation pulse had to be determined and a spike pulse was selected due to its broadband characteristics. The broadband type ultrasonic transducers have been widely used for various NDE applications because they can produce high-resolution characteristics for measuring thickness, sizing of defects, ultrasonic imaging techniques, and so on [27,28]. Figure 7 represents the excitation pulse used in this study.

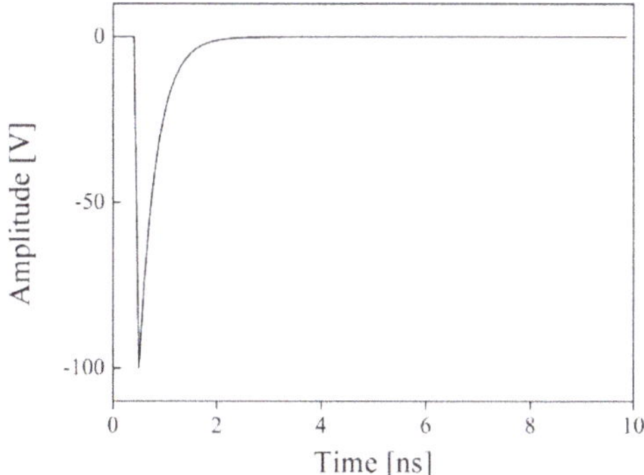

Figure 7. Simulated excitation pulse used in the KLM model modified for the through-transmission method.

In Figure 7, the delay and rise time of the excitation pulse was 0.2 and 2 ns, respectively, and the pulse voltage and relative bandwidth of the excitation pulse were −100 V and 150 MHz at −3 dB. The modified KLM simulation conditions are summarized in Table 3. The raw data of excitation pulse is included in Supplementary file 2.

Table 3. Simulation conditions of the KLM model.

Simulation Conditions	Units	Values
Area of PMN-PT	mm^2	9
Resonance frequency of PMN-PT	MHz	5
Sound velocity of front matching layer	m/s	2800
Thickness of the test material	mm	14
Acoustic impedance of water		1.5
Acoustic impedance of front matching layer		3.5×10^6
Acoustic impedance of backing material	$kg/m^2 s$	8×10^6
Acoustic impedance of bonding material		3×10^6

In Table 3, the acoustic impedance of the front matching layer was derived using Equation (1) and the acoustic impedance of the bonding material was the same as that of the epoxy resin. Based on these design conditions, the through-transmitted waveforms were simulated and the prototype ultrasonic transducer for NDE of PWR fuel rods was fabricated.

4. Results and Discussion

4.1. Prototype Ultrasonic Transducer for NDE of PWR Fuel Rods

The prototype ultrasonic transducer for NDE of PWR fuel rods was fabricated under the same conditions used in the modified KLM simulation and is represented in Figure 8.

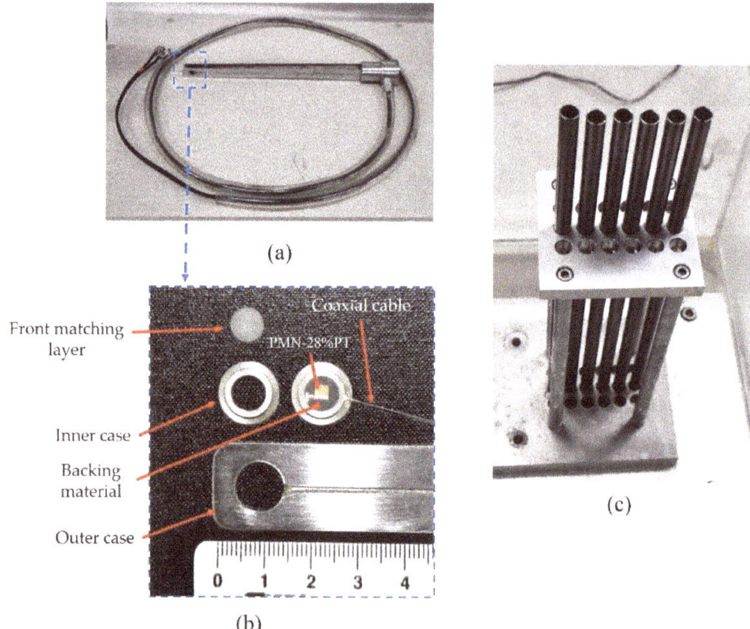

Figure 8. Newly developed PMN-28%PT-based ultrasonic transducer for NDE of PWR fuel rods: (**a**) Prototype ultrasonic transducer; (**b**) magnified image of ultrasonic transducer and its components; (**c**) PWR fuel rod assembly samples.

In Figure 8a,b, it can be seen that the newly developed ultrasonic transducer is composed of thin coaxial cable (≤0.3 mm), poly methyl methacrylate (a front matching layer), backing material, long coaxial cables (1 m, 50 Ω), and the inner and outer cases (stainless steel). The outer shield and center conductor of the coaxial cable was connected to both surfaces of the PMN-28%PT by a flat silver wire (25 μm, California Fine Wire Company, USA) with conductive metal epoxy (TC-2707, 3M, USA). The inner case was also bonded to the front matching layer and to the backing material by epoxy resin (Araldite GY509, Huntsman, USA) as shown in Figure 2. In this study, the radioactive resistance test for the backing material was not performed. However, we applied the backing material in this study because it has been used to our previous research works for the last 10-year period [5,25,26] and to the best of our knowledge, it was not easy to find proper backing materials with radioactive resistance for NPPs applications. Therefore, its radioactive resistance test will be also performed as a future work. The outer case was connected to the inner case assembly. The ultrasonic transducer for NDE of PWR fuel rods is typically designed to be inserted among the PWR fuel rods. These are arranged in a greedy assembly to prevent nuclear fuel vibration and movement in all directions, as shown in Figure 8c. In Figure 8a,c, the thickness and the center-to-center distance among the PWR fuel rods was 1 mm and 14 mm, respectively. Therefore, the distance between Tx and Rx transducers was 12 mm.

To fabricate the attenuating material for the backing, 25 μm tungsten powder (W006016, Good Fellow, England) was mixed with epoxy resin. The attenuating material needs high density and ultrasonic propagation speed, and strong scattering effect because its acoustic impedance is the function of the density and sound speed [28]. Tungsten satisfies these material characteristics (19,300 kg/m^3 for the density and 5180 m/s for the sound velocity). Moreover, its powder particles can effectively scatter the ultrasonic waves as well as absorb the mechanical vibration generated from the backside of the PMN-28%PT. In the middle of stirring the mixture, a lot of air bubbles can flow into the backing material. These should be removed because they decrease the acoustic impedance and the scattering effects.

Therefore, a vacuum desiccator (VDC-31, Jeio Tech Co., Ltd., Daejeon, Korea) was used to eliminate the voids. Because the viscosity of the adhesive material should be low to mix with tungsten powder, the araldite epoxy resin was selected (about 5 Pas). Finally, the through-transmitted waveforms of normal and defective specimens were acquired using the newly developed ultrasonic transducer and compared with the simulation outcome.

4.2. Through-Transmitted Waveforms and Comparison with the Simulation Resutls

The experimental set-up of the through-transmission method includes the newly developed ultrasonic transducer for NDE of PWR fuel rods, an oscilloscope (Wave Runner 640ZI, Lecroy, Chestnut Ridge, NY, USA), a pulse generator and receiver (HIS2, Krautkramer, Tokyo, Japan), and the defective and normal specimens. The excitation signal of the pulse generator was a spike pulse, and its amplitude and bandwidth were −100 V and 150 MHz at −3 dB and the same excitation pulse was simulated in Figure 4. As mentioned above, the PWR fuel rods are filled with sintered uranium dioxide pellets and they emit various radioactive fission products. Therefore, in this study, a PWR fuel rod specimen with sintered uranium dioxide pellets was not used for safety reasons. To verify the through-transmitted response waveforms of the designed ultrasonic transducer, we obtained the through-transmitted response signals of an empty PWR fuel rod (a normal specimen) and a water-filled PWR fuel rod (a defective specimen). Although the PWR fuel rod was not packed with the uranium pellets, it was enough to compare the experimental outcome with the simulation results, because the conditions of the experimental set-up and simulation were the same. Figure 9 represents the through-transmitted waveforms of both specimens.

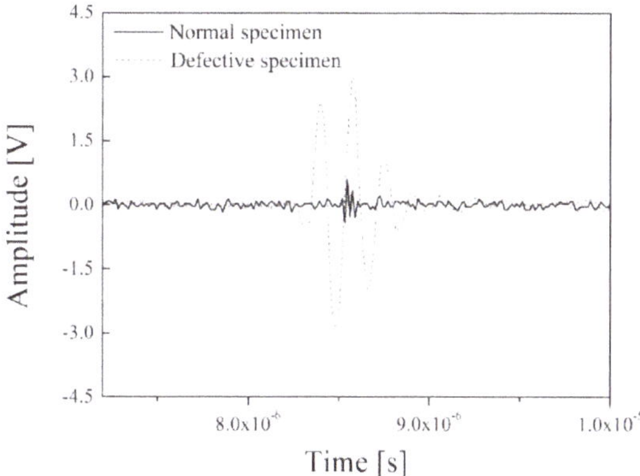

Figure 9. Through-transmitted response waveforms of the normal and defective specimens.

In Figure 9, the black line and gray dashes are the through-transmitted response waveforms of the normal specimen and the defective specimen, respectively. The peak-to-peak amplitudes of the normal specimen and defective specimen were about 0.9 and 5.7 V. As noted above, theoretically, the main path along which the ultrasonic energy is transmitted in the normal specimen is through a zircaloy tube described in Figure 4a. However, most of the transmitted signals can be trapped, reflected, and refracted in the tube due to its complex internal structure, cylindrical external shape, and relatively larger size than the sensing parts of the ultrasonic transducers, as shown in Figure 8a,b. In addition, because the acoustic impedance of each of the filling materials within the zircaloy tube is very different, the total acoustic energy transmission efficiency is very low. Therefore, those effects cause a large amount of acoustic energy loss during acoustic energy transfer. Although the normal PWR fuel rods are filled with the sintered uranium dioxide pellets, similar results should be derived in this case because the difference in the acoustic impedance of helium gas (0.00017×10^6 kg/m^{2}s) and a zircaloy tube (42.6×10^6 kg/m^{2}s) is very large. Thus, the transmitted acoustic waves are trapped in the tube due to the difference of acoustic impedance between the two materials and are reflected in various directions on the surface of the normal specimen. Therefore, the amplitude of the transmitted ultrasonic waveforms of the normal PWR fuel rods was very low or could be almost zero. From the above experimental results and analysis, simulation of the normal specimen was not conducted because its amplitude was too low to measure. In addition, the simulated through-transmitted response time of the defective PWR fuel rod was approximately 8.6×10^{-6}, which was considering the thickness of fuel rod (2 mm), permeated water (10 mm), and the distance between Tx and Rx ultrasonic transducers. However, we could not obtain the through-transmitted response time of the normal PWR fuel rod. In the case of ultrasonic inspection for normal PWR fuel rods, shown in Figures 1 and 5, three-dimensional cylindrical structures should be considered, and its acoustic energy would be trapped in round-shaped ring due to the great acoustic difference between helium has gap and PWR fuel rod. However, the KLM simulation cannot perform the three-dimensional acoustic wave propagation analysis and only two-dimensional analysis can be only performed. Therefore, acoustic energy is reflected on the helium gas layer and the simulated amplitude of normal PWR fuel rod was almost zero. To acquire more accurate results for the normal PWR fuel rods, we concluded that the additional study based on infinite element analysis is needed. Figure 10 shows the through-transmitted signals of the defective specimen and its simulation outcome.

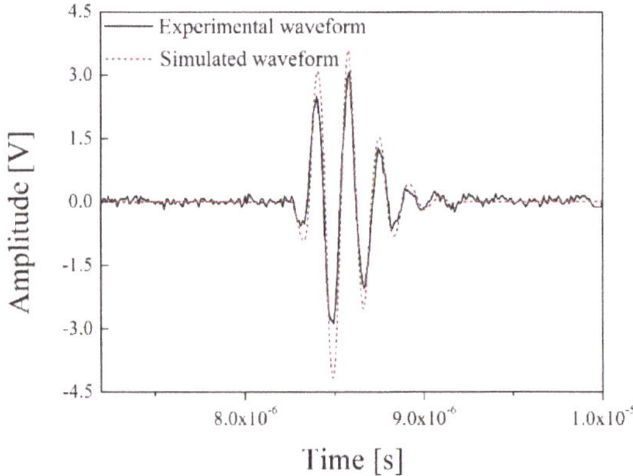

Figure 10. Through-transmitted waveforms of the defective specimen and its simulation result.

In Figure 10, the black line is the experimental through-transmitted waveform of the defective specimen and the red dashes describe its simulated waveform. The peak-to-peak amplitude of the simulated waveform is approximately 7.6 V, which is about 0.4 dB higher than its experimental result. In the simulation, the acoustic impedance of the defective specimen included only those of the tube and water. However, in the real world, the response waveform is affected by the spherical geometry of the zircaloy tube and it can cause acoustic energy loss and attenuation. Moreover, the electrical tuning needed to match the electrical impedance of the ultrasonic transducer to the ultrasonic system, including the oscilloscope and pulse generator, was not performed because the radioactive resistance of the electrical circuit and related effects were not investigated in this study. In addition, all electrical impedance in the simulation is ideally matched to 50 Ω. However, in our previous study, the electrical impedance of PMN-28%PT was changed with increase in the neutron radiation dose [16]. Therefore, it was not easy to design an optimal electrical tuning circuit due to its radioactivity dependence. In spite of the difficulties, the simulation results showed good agreement with the experimental results, only excepting the amplitude. Figure 11 represents the frequency spectrum (fast Fourier transform, FFT) of the two through-transmitted waveforms.

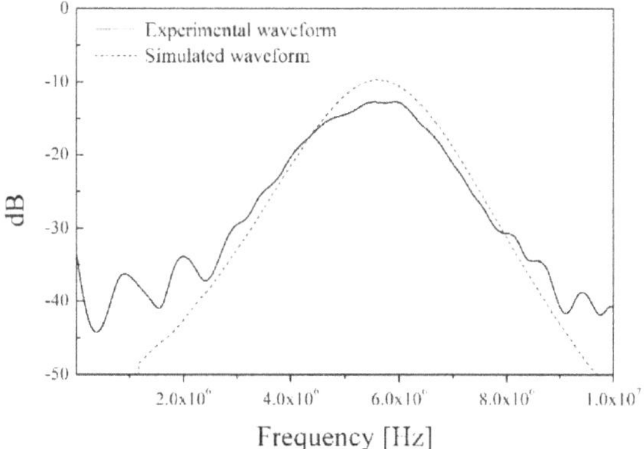

Figure 11. Frequency spectrum of through-transmitted waveforms of the defective specimen.

In Figure 11, the center frequency and bandwidth at −6 dB of two waveforms are almost same (about 5.5 MHz for the center frequency and about 40% for the bandwidth). In our previous study, backing material of about 8×10^6 kg/m²s was simulated and a PZT-based ultrasonic transducer with approximately 60% bandwidth was developed [25]. However, in this study, the bandwidth decreased to about 40% from 60% because of the high piezo-electric charge constant, d_{33}. PMN-PT produced high-sensitivity response signals due to its exceptional piezo-electric coefficient. However, its pulse duration was also increased because of its strong electro-mechanical conversion efficiency. The measured piezo-electric charge of the PMN-28%PT used was approximately 1515 pC/N, and this value is about three times higher than that of the normal PZTs widely used in the NDE field. From this point of view, the bandwidth is in inverse proportion to the piezo-electric charge constant, and the bandwidth of the newly developed ultrasonic transducer was also decreased. However, this is not a serious problem for detecting the infiltrated water in PWR rods. As shown in Figure 9, the clear difference between both through-transmitted waveforms of the normal and defective specimens was the amplitude of the through-transmitted waveforms. The bandwidth was not an important factor for inspecting the PWR fuel rods. Therefore, detecting the presence of infiltrated water should be conducted by analyzing the variation in the transducer amplitude. This section may be divided by subheadings. It should provide a concise and precise description of the experimental results, their interpretation, as well as the experimental conclusions that can be drawn.

5. Concluding Remarks

In this study, a high-sensitivity piezo-electric ultrasonic transducer for detecting water infiltrated PWR fuel rods was designed and fabricated. The substance PMN-28%PT was used as the piezo-electric material due to its high piezo-electric constants. The through-transmitted waveforms were simulated using a newly developed KLM model. Moreover, the experimental setup was constructed using the same conditions used in the simulation. The through-transmitted waveforms of normal and defective specimens were obtained and compared with the simulation results. The results clearly demonstrated an amplitude difference between through-transmitted waveforms of normal and defective tubes, and this was the key feature for detecting infiltrated water in PWR fuel rods. However, the through-transmitted response waveform of normal PWR fuel rod was not fully analyzed because of the limitation of KLM model. To analyze the ultrasonic behavior of three-dimensional cylindrical structures with big acoustic differences, a finite element analysis was essential. In addition, the simultaneous measuring system for neutron dose and temperature was required for more harsh condition of various reactors. Therefore, these additional studies should be conducted for our future work.

From the frequency spectrum analysis of the normal and defective waveforms, the bandwidth of the newly developed ultrasonic transducer was decreased because of its high piezo-electric coefficient.

However, this was not a crucial factor in distinguishing the defective PWR fuel rods. In conclusion, the newly constructed KLM model shows good agreement with the experimental results and the newly developed PMN-28%PT-based ultrasonic transducer should be a promising device for use in NDE of components in NPPs.

Supplementary Materials: The following are available online at http://www.mdpi.com/1424-8220/19/12/2662/s1, Supplementary file 1: Matlab codes for through-transmission method using Matlab software, Supplementary file 2: raw data of excitation pulse.

Author Contributions: Experiments, G.K. and N.C.; transducer design and fabrication, G.K.; simulation, G.K.; analysis, G.K., Y.-I.K., N.C. and K.-B.K.; writing, G.K.; supervision and funding acquisition, K.-B.K.

Funding: This research was supported by development of safety measurement technology for infrastructure industry funded by Korea Research Institute of Standards and Science (KRISS – 2019 – GP2019-0014).

Conflicts of Interest: The authors declare no conflict of interest.

References

1. Choi, M.S.; Yang, M.S.; Kim, H.C. Detection of leak-defective fuel rods using the circumferential Lamb waves excited by the resonance backscattering of ultrasonic pulses. *Ultrasonics* **1992**, *30*, 221–223. [CrossRef]
2. Womack, R.E.; Lawrie, W.E.; Jester, A. Apparatus and Method for Detecting Defective Fuel Rods. U.S. Patent No. 4,193,843, 18 March 1980.
3. Cross, H.D.; Hansen, L.E.; Mcclelland, R.G. Method and Apparatus for Ultrasonic Testing of Nuclear Fuel Rods Employing an Alignment Guide. U.S. Patent No 5,215,706, 1 June 1993.
4. Blackstone, E.G.; Lofy, R.A.; Williams, L.P. Nuclear Reactor Vessel Inspection Apparatus. U.S. Patent No. 4,169,758, 2 October 1979.
5. Kim, K.B.; Hsu, D.K.; Ahn, B.; Kim, Y.G.; Barnard, D.J. Fabrication and comparison of PMN-PT single crystal, PZT and PZT-based 1–3 composite ultrasonic transducers for NDE applications. *Ultrasonics* **2010**, *50*, 790–797. [CrossRef] [PubMed]
6. Kim, Y.I.; Kim, G.; Bae, Y.M.; Ryu, Y.H.; Jeong, K.J.; Oh, C.H.; Kim, K.B. Comparison of PMN-PT and PZN-PT single-crystal-based ultrasonic transducers for nondestructive evaluation applications. *Sens. Mater.* **2015**, *27*, 107–114.
7. Angadi, B.; Jali, V.M.; Lagare, M.T.; Bhat, V.V.; Umarji, A.M.; Kumar, R. Radiation resistance of PFN and PMN–PT relaxor ferroelectrics. *Radiat. Meas.* **2003**, *36*, 635–638. [CrossRef]
8. Cheng, K.C.; Chan, H.L.; Choy, C.L.; Yin, Q.; Luo, H.; Yin, Z. Single crystal PMN-0.33 PT/epoxy 1–3 composites for ultrasonic transducer applications. *IEEE Trans. Ultrason. Ferroelectr. Freq. Control* **2003**, *50*, 1177–1183. [CrossRef]
9. Sun, E.; Cao, W. Relaxor-based ferroelectric single crystals: Growth, domain engineering, characterization and applications. *Prog. Mater Sci.* **2014**, *65*, 124–210. [CrossRef]
10. Park, S.E.; Shrout, T.R. Characteristics of relaxor-based piezoelectric single crystals for ultrasonic transducers. *IEEE Trans. Ultrason. Ferroelectr. Freq. Control* **1997**, *44*, 1140–1147. [CrossRef]
11. Ritter, T. Single crystal pzn/pt-polymer composites for ultrasound transducer applications. *IEEE Trans. Ultrason. Ferroelectr. Freq. Control* **2000**, *47*, 792–800. [CrossRef]
12. Zhou, Q.; Xu, X.; Gottlieb, E.J.; Sun, L.; Cannata, J.M.; Ameri, H.; Humayun, M.S.; Han, P.; Shung, K.K. PMN-PT single crystal, high-frequency ultrasonic needle transducers for pulsed-wave Doppler application. *IEEE Trans. Ultrason. Ferroelectr. Freq. Control* **2007**, *54*, 668–675. [CrossRef]
13. Mustafaeva, S.N.; Asadov, M.M.; Ismaïlov, A.A. Effect of gamma irradiation on the dielectric properties and electrical conductivity of the TlInS 2 single crystal. *Phys. Solid State* **2009**, *51*, 2140–2143. [CrossRef]
14. Sheleg, A.U.; Iodkovskaya, K.V.; Kurilovich, N.F.; Pupkevich, P.A. Effect of gamma irradiation on the dielectric properties of γ crystals in the vicinity of the ferroelectric phase transition. *Neorg. Mater.* **2001**, *37*, 243–247. [CrossRef]
15. Medhi, N.; Nath, A.K. Gamma ray irradiation effects on the ferroelectric and piezoelectric properties of barium titanate ceramics. *J. Mater. Eng. Perform.* **2013**, *22*, 2716–2722. [CrossRef]
16. Kim, Y.I.; Choi, N.; Kim, G.; Lee, Y.H.; Baek, K.S.; Kim, K.B. Effect of neutron irradiation on properties of $Pb(Mg_{1/3}Nb_{2/3})O_3$–$PbTiO_3$. *J. Nanosci. Nanotechnol.* **2015**, *15*, 8414–8417. [CrossRef] [PubMed]

17. Alcala, R.F. *Application of Non-Destructive Testing and in-Service Inspection to Research Reactors*; IAEA-TECDOC-1263; International Atomic Energy Agency (IAEA): Vienna, Austria, 2001.
18. Available online: http://www.world-nuclear.org/information-library/current-and-futuregeneration/cooling-power-plants.aspx (accessed on 9 June 2019).
19. Yan, Y.; Cho, K.H.; Priya, S. Piezoelectric properties and temperature stability of Mn-doped Pb $(Mg_{1/3}Nb_{2/3})$-$PbZrO_3$-$PbTiO_3$ textured ceramics. *Appl. Phys. Lett.* **2012**, *100*, 132908. [CrossRef]
20. Redwood, M. Transient performance of a piezoelectric transducer. *J. Acoust. Soc. Am.* **1961**, *33*, 527–536. [CrossRef]
21. Mason, W.P. Design of electromechanical system. In *Electromechanical Transducers and Wave Filter*, 2nd ed.; D. Van Nostrand Company, Inc.: New York, NY, USA, 1948; pp. 225–238.
22. Krimholtz, R.; Leedom, D.A.; Matthaei, G.L. New equivalent circuits for elementary piezoelectric transducers. *Electron. Lett.* **1970**, *6*, 398–399. [CrossRef]
23. Castillo, M.; Acevedo, P.; Moreno, E. KLM model for lossy piezoelectric transducers. *Ultrasonics* **2003**, *41*, 671–679. [CrossRef]
24. Sherrit, S.; Leary, S.P.; Dolgin, B.P.; Bar-Cohen, Y. Comparison of the Mason and KLM equivalent circuits for piezoelectric resonators in the thickness mode. In Proceedings of the 1999 IEEE Ultrasonics Symposium. Proceedings. International Symposium, Caesars Tahoe, NV, USA, 17–20 October 1999.
25. Kim, G.; Seo, M.K.; Choi, N.; Baek, K.S.; Kim, K.B. Application of KLM model for an ultrasonic through-transmission method. *Int. J. Precis. Eng. Manuf.* **2019**, *20*, 383–393. [CrossRef]
26. Kim, G.; Seo, M.K.; Choi, N.; Kim, Y.I.; Kim, K.B. Comparison of PZT, PZT based 1–3 composite and PMN–PT acoustic emission sensors for glass fiber reinforced plastics. *Int. J. Precis. Eng. Manuf.* **2019**, 1–9. [CrossRef]
27. Kossoff, G. The effects of backing and matching on the performance of piezoelectric ceramic transducers. *IEEE Trans. Sonics Ultrason.* **1966**, *13*, 20–30. [CrossRef]
28. Desilets, C.S.; Fraser, J.D.; Kino, G.S. The design of efficient broad-band piezoelectric transducers. *IEEE Trans. Sonics Ultrason.* **1978**, *25*, 115–125. [CrossRef]
29. Petersen, G. *L-Matching the Output of a RITEC Gated Amplifier to an Arbitrary Load*; RITEC Inc.: Warwick, RI, USA, 2006.

MDPI

St. Alban-Anlage 66

4052 Basel

Switzerland

Tel. +41 61 683 77 34

Fax +41 61 302 89 18

www.mdpi.com

Sensors Editorial Office

E-mail: sensors@mdpi.com

www.mdpi.com/journal/sensors